Modeling Invasive Alien Plant Species

Fuzzy-Based Uncertainty

H.O.W. Peiris
S. Chakraverty
S.S.N. Perera
S.M.W. Ranwala

CRC Press
Taylor & Francis Group
Boca Raton London New York

CRC Press is an imprint of the
Taylor & Francis Group, an informa business

First edition published 2022
by CRC Press
6000 Broken Sound Parkway NW, Suite 300, Boca Raton, FL 33487-2742

and by CRC Press
2 Park Square, Milton Park, Abingdon, Oxon, OX14 4RN

Library of Congress Cataloging-in-Publication Data

Names: Peiris, Habaragamuralalage Osadee Widarshi, author.
Title: Modeling invasive alien plant species : fuzzy-based uncertainty / Habaragamuralalage Osadee Widarshi Peiris, Snehashish Chakraverty, Shyam Sanjeewa Nishantha Perera, Sudheera Manorama Wadisinh Ranwala.
Description: First edition. | Boca Raton : CRC Press, 2021. | Includes index.
Identifiers: LCCN 2021006018 (print) | LCCN 2021006019 (ebook) | ISBN 9780367758097 (hardback) | ISBN 9781003193807 (ebook)
Subjects: LCSH: Invasive plants--Mathematical models. | Invasive plants--Risk assessment. | Fuzzy mathematics.
Classification: LCC SB613.5 .P45 2021 (print) | LCC SB613.5 (ebook) | DDC 581.6/2--dc23
LC record available at https://lccn.loc.gov/2021006018
LC ebook record available at https://lccn.loc.gov/2021006019

ISBN: 978-0-367-75809-7 (hbk)
ISBN: 978-1-032-04564-1 (pbk)
ISBN: 978-1-003-19380-7 (ebk)

DOI: 10.1201/9781003193807

Typeset in Times
by KnowledgeWorks Global Ltd.

Contents

List of Figures

List of Tables

Preface

The era in which we live, is full of complexities and ambiguities even with technological advancements. The problems that we need to solve are not always defined clearly and are accompanied by uncertainties, impreciseness, and vagueness. That means most of the time we need to deal with fuzziness. These problems may arise from the relationships between human and environment/machines. When solving these kinds of relationships, people often face difficulties such as that the accompanied parameters may not be easily handled and quantifiable, though such problems require a sophisticated scientific treatment. Classical logic and conventional statistical techniques are not always able to handle such complex systems. In contrast, fuzzy set theory shows the capability of handling such an uncertain and imprecise environment where it works with partial truth.

The main aim of this book is to provide comprehensive modeling techniques within a fuzzy set theory framework to facilitate solving problems arising from an uncertain, imprecise, and vague environment. It gives a guide for systematic approaches for developing models using various techniques in fuzzy set theory and provides evidence that these models may be successfully applied to evaluate the risk of invasive plant species.

This book is a multidisciplinary text, and concerning the coverage we hope that it is a good reference for postgraduates and researchers in various areas of engineering, science, finance, management, etc. in which decision making is prominent. It presents many examples, figures, and tables so the reader can grasp the knowledge of basics of fuzzy set theory easily; the way to handle uncertain and imprecise parameters and how these models can be applied to a real situation are also clearly explained.

This book consists of eleven chapters each divided into sections. The references and an index are in the end of book.

Chapter 1 addresses the overview of invasive alien plant species. Their impact and threats are discussed thoroughly. Also, how these species become invasive is pointed out.

Chapter 2 discusses mainly the role of risk assessment in investigating hazards, and discusses present risk assessments available for evaluating invasive plant species and their drawbacks. At the end, it elaborates upon the importance of modeling the risk of invasive plant species via fuzzy set theory.

Chapter 3 begins with a brief review of the development of fuzzy set theory. It then provides a basic idea of classical sets and clearly explains the how a fuzzy set is different from a classical set. The important properties of fuzzy sets, membership functions, and different types of membership functions are introduced. Basic fuzzy set theoretic operators and averaging operators are also presented.

Chapter 4 focuses on modeling quantitative parameters. Two systematic approaches of developing factor-based models for problems accompanied with only quantitative parameters are given. It begins with explaining how to define the fuzzy set and choose its appropriate membership function for an uncertain and imprecise parameter. The first approach deals with incorporating importance weights of the parameters viz. concentration/dilation operators. The second approach reveals the importance of incorporating the experts' opinions' in decision making by using fuzzy analytical hierarchy techniques.

Chapter 5 presents three approaches based on interval multiple linear regression techniques which may be used to model both quantitative and qualitative parameters. The first half of the chapter discusses the basic preliminaries which are relevant to this work. To find the interval coefficients, three approaches are given. The first approach is based on the interval least squares algorithm proposed by Chenyi Hu. The other two approaches are based on the novel methods.

Chapter 6 covers providing factor-based models appropriate for the parameters which are qualitative/linguistic in nature. The work describes two approaches. The first approach is the symbolic approach which directly computes the labels of the parameters and consists of two methods where the first and second methods are based on fuzzy linguistic operators and the fuzzy 2-tuple linguistic approach, respectively. In the first method three models are given based on the linguistic operators: Linguistic Ordered Weighted Average, Linguistic Weighted Average, and Majority Guided Induced Ordered Weighted Average. The second method incorporates non-weighted and weighted operators which are defined for the 2-tuple linguistic approach. The second approach is based on the approximation which considers the semantics of the labels. A merging method which combines two labels from two different linguistic scales is discussed. A model structure has been introduced where any situation is accompanied by different scales for importance weights and the performance values of the parameters by incorporating the merging method.

Chapter 7 presents a systematic approach to building a factor-based model which is accompanied by an unbalanced linguistic terms set. In the first half of the chapter, a mechanism for assigning the semantic of labels in an arbitrary unbalanced scale with seven linguistic labels is explained step by step. The way to handle bridging labels which represents two levels in the linguistic hierarchy and assigns their semantics is also discussed. In the latter part, the structure for a factor-based model accompanied by an unbalanced scale is given. The operators available in the unbalanced linguistic approach to aggregate the parameters are also discussed.

Chapter 8 discusses the application of models present in Chapter 4 to evaluate the invasion risk of invasive alien plant species. In the first half of the chapter, the process of establishing membership functions for the identified parameters and their modification is discussed. The chapter presents four models based on fuzzy set theory operators: Hamacher, Yager, Dubois and Prade, and Dombi, and their validation results are also given. The discussion on output of the four models and their performance comparisons are also given. It also presents two models each from methods relating to the fuzzy analytical hierarchy process. The test and validation results of these models and a discussion are also given.

Chapter 9 discusses the application of models present in Chapter 5 to evaluate the invasion risk of invasive alien plant species. It presents three models which consist of twelve parameters relating to invasiveness, and the validation results of the models are given. The quality measurement of each model and a brief discussion are also given.

Chapter 10 presents several models obtained from model structures given in Chapter 6 to evaluate the invasion risk of Invasive Alien Plant Species (IAPS). Three models based on three linguistic operators are presented and their test and validation results are also given. Two models are given from the fuzzy 2-tuple approach with test and validation results. Two models are given from the structure viz approximation approach test and validation results. A brief discussion on each model is also provided.

Chapter 11 discusses developing models based on the model structures given in Chapter 7 to evaluate the invasion risk of Invasive Alien Plant Species (IAPS). In the first half of the chapter, the process of assigning the semantics of the seven labels in the unbalanced scale using an appropriate linguistic hierarchy is clearly explained. The second half of the chapter discusses the two models built upon the unbalanced scale. The test and validation results of these models are also given with a brief discussion.

Acknowledgments

We do greatly appreciate all the researchers and authors mentioned in the references of this book who have directly or indirectly helped us to write this book in a systematic way.

The first author expresses sincere gratitude for her parents and sister for their continuous support and love. She greatly appreciates her husband for his enduring love and for sharing her wish to reach the goal of completing this task. Further, she would like to thank her daughter Yesali for being patient during this period.

The second author would like to thank his parents, the late Sri B. K. Chakraborty and the late Sri Mati Parul Chakraborty for their blessings. Further, he would like to thank and appreciate his wife Shewli and daughters Shreyati and Susprihaa for their love, support, and source of inspiration throughout.

All the authors would also like to appreciate the University of Colombo and NIT Rourkela authorities for moral and other support.

The fruitful comments and suggestions of the reviewers are sincerely acknowledged. Heartfelt thanks are also due to all the scientists who have contributed to various extents to develop the theories and concepts in fuzzy set theory which provided the authors baseline information to build up this research study.

Finally, the authors do appreciate the whole team of CRC Press, Taylor & Francis Group, who supported us throughout this project to ensure timely publication.

H.O.W. Peiris
S. Chakraverty
S.S.N. Perera
S.M.W. Ranwala

1 Invasive Alien Plant Species

Increased travel, trade, and tourism associated with globalization and expansion of the human population have facilitated intentional and unintentional movement of many plant and animal species, and of microorganisms beyond the world's natural bio-geographical barriers. Few of these species have being "troublesome" to control by mankind and have continued to invade new regions at an unprecedented rate, replacing the native fauna and flora of the new environment. Invasive alien species are therefore recognized as one of the drivers of global biodiversity change due to the adverse ecological and economical impacts they impose on the biodiversity of introduced environments (Veich and Clout 2002, Pimental et al. 2001, Vitousek, D'Antonio, et al. 1997).

The term Invasive Alien Species (IAS) is defined as non-indigenous (non-native or exotic) species of a plant, animal, or micro-organism, that have been deliberately or accidentally introduced to new areas beyond their native range and which then spread beyond human management impacting the biodiversity in the new environment (CBD 1994).

Richardson et al. (2000) provided a detail review on concepts involving and definition of Invasive Alien Species. Invasive Alien Species have also been identified by scientists from the human perspective considering the negative impacts caused by IAS. For example, Arbaciauskas et al. (2008) identified biological invasions as "biocontamination" because of the spoiled nature of the environment during invasion. Johns, Lawton, and Shachak (1994) considered IAS as "ecosystem engineers" because they are capable of altering ecosystem goods and services. It is also believed that microscopic invaders, especially in aquatic habitats, can be considered as "bioindicators" as they reflect environment health either directly or indirectly. IAS therefore are "silent killers" and/or a kind of "bioweapon" (Zaghloul, et al., 2020, Parker et al., 1999).

1.1 INVASIVE ALIEN PLANT SPECIES (IAPS) AND THEIR IMPACT ON THE ENVIRONMENT

Invasive Alien Plant Species (IAPS) could be herbs, shrubs, creepers, and/or trees, found in both terrestrial and aquatic environments. IAPS can invade almost every ecosystem on earth consuming excessive amounts of resources (notably nutrients, water, light, oxygen, and living space for other organisms) due to their efficiency in resource use, including greater ability to capture light leading to higher rates of photosynthesis, fast growth, and greater fruit production. Hence, they have the ability to threaten the community structure, function, and productivity of natural ecosystems (Pyšek, Jarošík et al., 2012; Vilá, Espinar, et al., 2011; Cronk and Fuller 1995).

However, impacts could be greatly variable on the landscape context and conditions of the ecosystems (Stromberg et al., 2007; Alpert, Bone, and Holzapfel, 2000; Davis, Grime, Thompson, 2000; Higgins, Richardson and Cowling, 1999).

Invasive Alien Plant Species change ecosystem structure, ecological relationships between native species, and ecosystem function. For example, in aquatic ecosystems, mat-forming populations of floating invasive plant species Water Hyacinth (Eichhornia crassipes) and submerged Waterthyme (Hydrilla verticillate) directly reduce penetration of light into the water and change water chemistry and oxygen levels threatening the biodiversity of the aquatic environment (Aloo et al., 2013; Langeland, 1996). In land environments excessive shading caused by monospecific stands of plant invaders such as shrubby Simpoh Air (Dillenia suffruticosa), Lantana (Lantana camara), and woody Mesquite (Prosopis juliflora) have greatly suppressed the undergrowth via preventing seed germination, establishment of seedlings, and inhibiting plant growth and development. IAPS also impact soil physical, chemical, and biological properties (Wickramathilake, Weerasinghe, and Ranwala, 2013; El-Keblawy and Al-Rawai, 2007; Gooden, et al., 2009; Richardson, Macdonald, and Forsyth, 1989). They transform soil properties by promoting or suppressing fire, stabilizing or destabilizing sand movement and/or erosion, accumulating litter, and redistributing minerals nutrient flow (Souza-Alonso, Novoa, and González, 2014; Lindsay and French, 2006; Brooks, 2008). In many situations IAPS increase above and below ground competition (Wang, et al., 2017). Due to global climate change, landuse changes, overgrazing, and localized nutrient enrichment through agrochemical usage, the competitive rate of IAPS is enhanced [Mack et al., 2000; Bradley, et al., 2010]. Some IAPS do interfere with native pollinators and modify plant-pollinator interactions (Bartomeus, Fründ, and Williams, 2015), bring about changes in animal behavior due to the presence of injurious hairs, spines, and toxin production (Bjerknes, et al., 2007) IAPS also swamp native gene pools via interbreeding with native species (Ellstrand and Kristin, 2000). Moreover, aggregate effects of multiple invasive species can cause large and complex damage to ecosystem properties, structure, and function ultimately affecting the goods and services provided by the environment for human well being.

In terms of economics, the costs of invasive alien species are significant as they cause significant economic losses to the government, industrial sector, and private citizens (Meyerson and Reaser, 2002) and in many countries agriculture, fisheries, and aquaculture, forestry, health sectors and conservation sectors have been most affected. Apart from these, the "ecotourism" sector is also badly affected due to the downgrading of recreational and cultural heritage values caused by uncontrollable spread of IAPS. Total annual costs, including losses to crops, pastures, and forests, as well as environmental damage and control costs, have been conservatively estimated to be a large amount in many countries. A study conducted by Pimentel, Lach, Zuniga, Morrison (2000) estimated that in the United States of America IAS had cost more than U.S. $100 billion a year. For an example the IAPS, Hydrilla, infestation in just two lakes in Florida were reported to cost $10 million dollars annually (Center, Frank et al., 1997). In South Africa, the specific cost of management of an aquatic species was reported to be U.S. $48 million spent on clearing invasive trees due to their impact on surface runoff (Pimentel, et al., 2000). Nevertheless, studies on financial impacts of IAPS are rather scarce due to the lack of information and limitations in maintenance of records for management of individual species (Iqbal, Wijesundera, and Ranwala, 2014). Hence considerable uncertainty remains regarding the estimated cost of managing invasive species.

Other than the ecological and economic aspects, IAPS pose great health risks to humans and other species by causing health related issues as the IAPS themselves could act as pathogens and parasites or serve as hosts accommodating the vectors (Meyerson and Reaser, 2002).

1.2 HOW CAN AN ALIEN SPECIES BECOME INVASIVE?

It is very important to become aware of how a non-native species become invasive. Predicting which plants will become invasive is a challenge. According to Richardson et al. (2000) the invasive pathway can be categorized into four major phases; (i) Introduction, during which an alien species travels from its native region to new habitats; (ii) colonization, when species tolerate, survive, and are fertile for longer periods in the introduced environments; (iii) establishment, the period of successful settlement in the habitat acquiring critical resources and resisting enemies; and (iv) spread, during which alien organisms spread and acquire the status of an invader in the habitat to which they have been introduced. Hence, it is important to consider that certain plant species show invasive behavior after a considerable period of time since their first introduction. In many instances activities associated with socio-economic development, including the 3T's–Trade, Travel and Tourism–have been identified as operational pathways or routes by which IAS spread in space and time. The ornamental plant trade is an example of a key route through which many species have been distributed beyond their home range. Unwillingly a large proportion of them have become invasive especially when their establishment and spread is favored in the introduced environment (Cronk and Fuller, 1995; Mack et al., 2000; Richardson et al., 2000). Invasive species tend to establish easily as natural enemies are unavailable immediately after their introduction into the new environment. Inherent invasive characteristics of IAPS make them superior competitors over native flora strengthening their successful establishment (Kolar and Lodge, 2001; Pyšek and Richardson, 2007).

Baker 1965 first described that the traits related to invasiveness are the following: easy germination of seeds, great longevity of seeds, rapid growth through vegetation phase to flowering phase, continuous seed production for as long as growing conditions permit, cross pollination, high seed output, adaptations for both short and long distance dispersal, vigorous vegetative reproduction of regeneration from a fragment, and ability to compete successfully with neighbors. Subsequently studies have identified that traits related to reproductive potential, vegetative reproduction, and dispersal are as important drivers of invasiveness (Pyšek and Richardson, 2007; Rejmanek and Richardson, 1996). According to Pyšek et al. (2009), traits related to reproduction such as early flowering and opportunistic dispersal over long distances by agents such as water, wind, vehicles, or livestock (often at remarkably high speeds) play an important role in the establishment and spread of IAPS. Hence, it is well noted that invasion success is strongly influenced by the production, dispersal, and genetic constitution of propagules. In addition, human-aid disturbances to ecosystems not only signify an extensive propagule pressure but also diminish the ecosystem's resistance inviting and encouraging invasive species to establish well in many environments (Gallardo and Aldridge, 2013).

1.3 MANAGEMENT OF INVASIVE ALIEN PLANT SPECIES

In general, a three-stage hierarchical strategic approach to minimize the risk and the spread of IAS is suggested. As prevention is better than control, all possible steps should be taken to prevent the introduction of IAS. Prevention restricts the introduction of any reproductive fragments of IAPS. However, at this stage correct identification of IAPS is essential. If prevention had failed vigilant early detection and eradication by the society plays a vital role in control and management of IAS. If that stage has already been passed, suitable control measures are required. Containment also allows the populations to be managed at a low level without further spread.

For successful control of a population of an invasive species, mechanical, chemical, and biological control methods can be used. Physical removal of IAPS is conducted with or without the use of machinery. However, this process consumes more human labor. Chemical control is quite expensive and typically requires repeat applications. However, it may result in the development of resistance in IAPS and may also harm non-targeted species. A biological control method includes an introduction of a highly specific predator, parasite, or pathogen that will attack the particular IAPS. This process does not totally remove the organism but can reduce the population. The initial costs are generally high due to research and development methods adopted to recommend the exact biocontrol agent, but the long-term cost is relatively low.

Currently, alternate management streams exploiting IAPS as a resource are also being implemented in some countries, thereby reducing the management costs. Kannan, et al. (2016) explored the possibility of using the invasive species Lantana camara as a substitute for bamboo and canes for the weaving purposes of the rural forest dwelling communities in India such that it could enhance the livelihoods of the traditional weaving communities who suffered from the scarcity of bamboo resources.

Nevertheless, some species might gain a benefit from invasive species. Sometimes invasive species may provide food and shelter for other native species (Thomsen, 2010) as many native animals use invasive plant species as habitat; thus in such cases the diversity and abundance in their populations can reach higher levels in habitats with invasive plants when compared with uninvaded habitats.

2 Risk Assessment of Invasive Alien Plant Species

2.1 RISK ASSESSMENT(RA) – IAPS

Risk assessments of IAPS provide a foundation to quantify the threat by IAPS on the environment based on available information and thereby could be used to identify the most troublesome species to prioritize IAPS management efforts. Risk assessment of IAS therefore is used to support the exclusion of potential invasive species from being introduced, as well as to assess the impacts of those that have already become established in a particular region. Hence, risk assessment is also considered as an early warning system to screen proposed imports and specify the likelihood of a species becoming invasive, as well as to evaluate weed incursions already found in the region to prioritize control programs.

Baker's (1965) list of weed characteristics is reported as an important initiative that paved the foundation for research in weed ecology and the development of weed risk assessments. The concept and use of weed risk assessments to evaluate weed potential developed in the 1980s with the evaluation of plant imports using a systematic science-based process prior to importation. In 1988 a scoring system was first implemented in Australia which later led to the development of the Australian WRA system (Koop, 2012). The Australian Weed Risk Assessment system (WRA) modified by Pheloung, Williams, and Halloy (1999) included an assessment of aggressiveness of species based on the success or the failure at different stages of the invasion process. During development of the Weed Risk Assessments various aspects were added, for example, the significance of the impact of suspected species and their feasibility of control or management (Hiebert and Stubbendieck, 1993).

A risk assessment procedure to regulate the introduction of invasive species in Sri Lanka first developed by Ranwala et al. (2012) includes a set of questions and scores related to the ecological impacts caused by plants, their invasive potential, and the current distribution and feasibility of control. The validation of the questions and respected scores were based on IAPS and non-IAPS and the scoring for known IAPS was conducted through several workshops of stakeholder consultation.

The updated National Risk Assessment (NRA) consisted of three main sections: i) current distribution of the IAS, ii) ecological and economical impacts of IAS on ecosystems, and iii) invasive attributes of the species. Each section was addressed in detail to incorporate several important related aspects in the form of questions. The questions were developed to be simple and user friendly with minimum usage of scientific terms. The cover page provided instructions for the evaluator while the last page of the assessment provided some explanatory notes for easy reference. The multiple answers provided an opportunity for the evaluator to identify the most suitable answer and its respective score that could be assigned to the species for each question. Together with the risk assessment a score sheet was provided. UNDP/GEF funded the project on Strengthening Capacity to Control the Introduction and Spread of Invasive Alien Species (IAS) in Sri Lanka which established the National Risk Assessment in Sri Lanka.

Basically, almost all risk assessment procedures focus more or less on the same aspects, for example, the range of distribution plant species, impacts to the economy, ecological importance, and assessment of invasive characteristics. Nevertheless, it is important to note that the evaluator should have a good knowledge of the subject for assessing scores and calculation of the final risk score for a test species. As a rule of thumb, the evaluator is advised to search and use information from trusted sources, preferably published data. In this situation the evaluator may need to carefully extract the most suitable and appropriate information from the pool of knowledge available. However, the evaluator may be faced with situations where there is no availability of any trustworthy recorded data on a particular aspect. At this level, careful observations of stakeholders directly involved in the sector may assist.

Hence, during risk assessments it will be worthwhile to understand the precision of the answers provided by the evaluator. This can be accompanied with an uncertainty assessment.

2.2 MODELING RISKS OF INVASIVE ALIEN PLANT SPECIES

One may note that for some questions, information is lacking or is very difficult to obtain for most of the IAPS even if there is clear evidence of their importance in successful invasions, and data are gathered as the opinions from group of experts. The most opinions may be contributed by the experts using their knowledge and field experiences. As such, knowledge is widely used in those risk assessments due to a lack of peer reviewed or other documented information on their introduction pathways, impacts, and possibilities for management. The expert knowledge is defined as substantive information on a particular topic that is not widely known by others and expert judgment as predictions by experts of what may happen in a particular context (Martin, et al., 2013). While expert judgment is useful and practical, it also leads to making the assessment environment imprecise and uncertain.

From a mathematical point of view, imprecision and uncertainty of those factors in the assessments should be handled well to obtain precise information on species. Moreover the final score obtained from RA's depends on the user; i.e., user bias may affect the final risk value of a particular species, and in many cases, it has been a manual process that takes a considerable time to complete. Also if the final output of the RA is in the form of risk level rather than a numerical value such an assessment is more informative. We believe that integration of a mathematical approach with risk assessments may assist (at least partly) to overcome this situation and increase the efficacy of the process.

We emphasize that integration of mathematical techniques is imperative to develop a comprehensive risk analysis. Among the existing techniques, fuzzy set theory can handle impreciseness and uncertain situations rather than statistical tools because many cases accompanied with uncertainty cannot be solved using probability theories. Human perception of words may be different from person to person. Therefore, aspects of fuzzy set theory are advantageous to capture the uncertainty of words used in the RA.

3 Preliminaries

3.1 AN OVERVIEW

Over the past two decades, there has been a rapid growth in the number and variety of applications in fuzzy sets and theory. Applications of this theory can be found in many disciplines such as engineering, medicine, meteorology, artificial intelligence, decision theory, experts system, management science, etc. Mathematician Lotifi Zadeh first introduced fuzzy sets, which emerged from a close gap between mathematics and the intuitive way that humans talk, think and interact with the world. Several books related to this area have also been written by different authors such as Zimmermann (2001); Lee (2005); Lin, Cao, and Liao (2018); Harris (2006); Bojadziev and Bojadziev (2007); Berkan and Trubatch (2000); Yager and Zadeh (1992); Kandel (1992); Dubois and Prade (1980a); Fodor and Roubens (1994); Kacprzyk and Fedrizzi (1990); Kickert (1979); Orlovsky (1994); Lowen (1992); Chen and Hwang (1992).

Fuzzy set theory can be viewed as an extension of classical set theory concepts. A classical set is defined by crisp boundaries where the universe of discourse is split into two groups: members and non-members, for example, the set of integers 1 to 10. Therefore there is a sharp boundary between members of the set and those not in the set. But there are some sets which do not have sharp boundaries (e.g. the set of tall people.) Therefore classical sets cannot be used to work sets with unsharp boundaries.

Unlike classical set theory, fuzzy set theory focuses on the degree of being a member of a set. The term "fuzziness" primarily describes uncertainty or partial truth (Berkan and Trubatch, 2000). Some of the similar terms related to this term are imprecision, ambiguity, vagueness, undecidedness. Fuzziness occurs when a piece of information is not clear cut. Fuzzy set is a set with unsharp and vague boundaries. In the following sections, we present some important concepts, notations, and definitions of fuzzy set theory.

3.2 FUZZY SETS

Classical Set

A classical set is a set with a crisp boundary. An element in this set either belongs to the set or not. It can be mathematically represented by the characteristic function. If the set under investigation is A, testing of an element x using characteristic function χ is expressed as (Zimmermann, 2001)

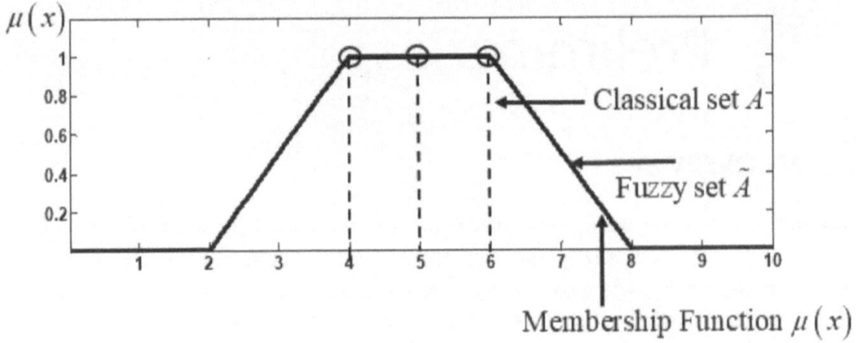

Figure 3.1 Representation of membership functions of classical sets and fuzzy set. (Mathe-works, 2020.)

$$\chi_A(x) = \begin{cases} 1 & x \in A \\ 0 & x \end{cases} \tag{3.1}$$

For example, a classical set A of real numbers greater than 7 can be expressed as

$$A = \{x | x > 7\} \tag{3.2}$$

where there is a clear, certain boundary 7. If x is greater than 7, then x belongs to the set A; otherwise x does not belong to the set.

Definition 3.1. (Zimmermann, 2001) If X is a collection of objects generically by x, then a fuzzy set $\tilde{A}$ in X is a set of ordered pairs:

$$\tilde{A} = \{(x, \mu_{\tilde{A}}(x)) | x \in X\}. \tag{3.3}$$

$\mu_{\tilde{A}}(x)$ is called membership function or grade of membership or degree of compatibility of x in $\tilde{A}$. The $\mu_{\tilde{A}}(x)$ maps each element of X to a membership value between 0 and 1.

3.3 MEMBERSHIP FUNCTION

Membership function fully defines the fuzzy set which is the major component of fuzzy set theory and it enables the fuzzy approach to access uncertain and ambiguous matters (Yager and Zadeh, 1992; Zhao and Bose, 2002). The membership function that indicates a fuzzy set $\tilde{A}$ is usually denoted by $\mu_{\tilde{A}}$. For an element x of X where the

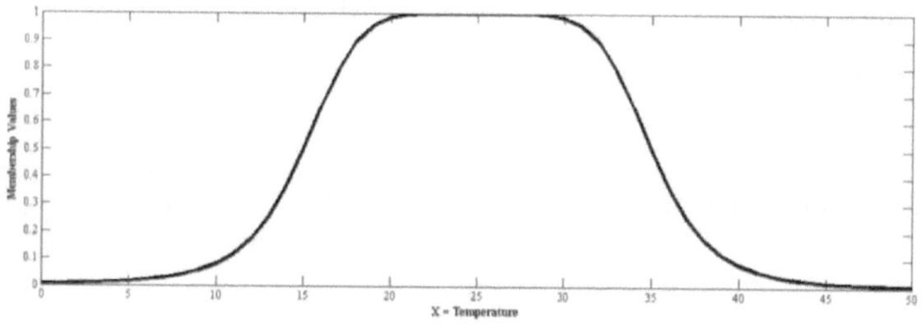

Figure 3.2 Membership function on a continuous universe. (Matheworks, 2020.)

X denotes the universal set (crisp), the value $\mu_{\tilde{A}(x)}$ is called the membership degree of x in the fuzzy set $\tilde{A}$. Some elements of a fuzzy set may have degree of membership zero since we generally consider the universal (crisp) set. Figure 3.1 depicts the difference between classical sets and fuzzy sets examined by the difference between two membership functions.

Example 3.1 *Fuzzy sets with a discrete universe*
Let $X = \{\text{Canada, United States, United Kingdom, France, Australia}\}$ be the set of countries to which one may like to immigrate. The set $\tilde{A}$="desirable country to immigrate" may be described as follows:
$\tilde{A} = \{(\text{Canada}, 0.8), (\text{United States}, 0.88), (\text{United Kingdom}, 0.65), (\text{France}, 0.6), (\text{Australia}, 0.7)\}$.
As one may see, the universe of discourse contains non-ordered objects and the membership values listed above are subjective, which may vary from person to person.

Example 3.2 *Fuzzy sets with a continuous universe* Let $X = \Re^+$ be the set of possible temperatures in a particular environment. Then the fuzzy set $\tilde{B}$="Temperature closer to 25 degrees Centigrade" may be expressed as
$\tilde{B} = \{(x, \mu_B(x)) \,|x \in X\}$,
where

$$\mu_{\tilde{B}}(x) = \frac{1}{1 + \left(\frac{x-25}{10}\right)^4}.$$

The membership function for the fuzzy set $\tilde{B}$ is shown in Figure 3.2.

In the following subsection we define terminology as it appeared in the literature to describe the membership functions explicitly.

3.3.1 TERMINOLOGY OF MEMBERSHIP FUNCTIONS

(i) **Support:** (Zimmermann, 2001)
The support of a fuzzy set $\tilde{A}$ is the crisp set of all points $x \in X$ such that $\mu_{\tilde{A}(x)} > 0$.
It can be mathematically expressed as: $\text{Support}(\tilde{A}) = \{(x, \mu_{\tilde{A}(x)}) | \mu_{\tilde{A}(x)} > 0\}$.

(ii) **Core:**
The core of a fuzzy set $\tilde{A}$ is the crisp set of all $x \in X$ such that $\mu_{\tilde{A}(x)} = 1$, and it can be expressed as: $\text{core}(\tilde{A}) = \{x \in X | \mu_{\tilde{A}(x)} = 1\}$.

(iii) **Normality:** (Bojadziev, 2007)
A fuzzy set $\tilde{A}$ is normal if its core is non-empty; i.e. $\text{core}(\tilde{A}) \neq \emptyset \Rightarrow \tilde{A}$ is a normal fuzzy set.The mathematical expression is $\text{Normality}(\tilde{A}) = 1$ if $\mu_{\tilde{A}(x)} = 1$, for all $x \in X$ and $(x, \mu_{\tilde{A}(x)}) \in \tilde{A}$.

(iv) **α-cut/α-level:**(Zimmermann, 2001)
The crisp set of elements that belong to the fuzzy set $\tilde{A}$ at least to the degree α is called the $\alpha-$level set:

$$\tilde{A}_\alpha = \{x \in X | \mu_{\tilde{A}(x)} \geqslant \alpha\}.$$

Strong α-cut is defined as

$$'\tilde{A}_\alpha = \{x \in X | \mu_{\tilde{A}(x)} > \alpha\}.$$

(v) **Convexity of Fuzzy Sets:**(Zimmermann, 2001)
A fuzzy set $\tilde{A}$ is convex if and only if for any $x_1, x_2 \in X$ and there exists $\lambda \in [0,1]$ such that

$$\mu_{\tilde{A}}(\lambda x_1 + (1-\lambda)x_2) \geq \min\left(\mu_{\tilde{A}(x_1)}, \mu_{\tilde{A}(x_2)}\right)$$

Figure 3.3 illustrates the core, $\alpha-$cut, and support of a bell-shaped membership function.

3.3.2 TYPES OF MEMBERSHIP FUNCTIONS

Here we present some commonly used parameterized membership functions.

(i) **Triangular Membership Function** (Bojadziev, 2007)
A triangular membership function is defined by three parameters $a, b, c \in X$ as

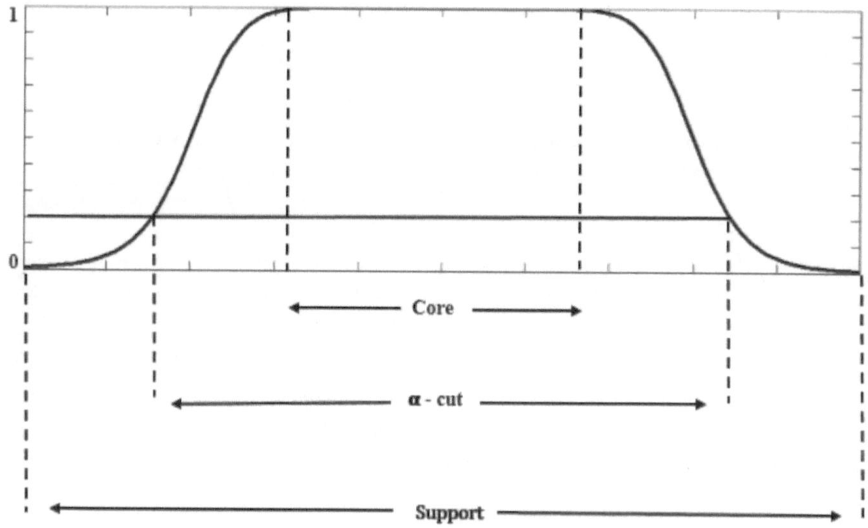

Figure 3.3 Core, α-cut, support of a fuzzy set. (Matheworks, 2020.)

follows:

$$\mu_{\tilde{A}}(x;a,b,c) = \begin{cases} 0, & x \leq a. \\ \frac{x-a}{b-a}, & a \leq x \leq b. \\ \frac{c-x}{c-b}, & b \leq x \leq c. \\ 0, & c \leq x. \end{cases} \tag{3.4}$$

The parameters $\{a,b,c\}$ (with $a < b < c$) represent the x coordinates of the three corners of the underlying triangular membership function (a: lower boundary and c: upper boundary where membership degree is zero, b: the center where membership degree is 1). The triangular membership function for the parameters $a = 2$, $b = 4$ and $c = 6$ is depicted in Figure 3.4(a).

(ii) **Trapezoidal Membership Function** (Bojadziev, 2007)
The trapezoidal membership function is defined by four parameters $\{a,b,c,d\}$ where $a,b,c,d \in X$ as follows:

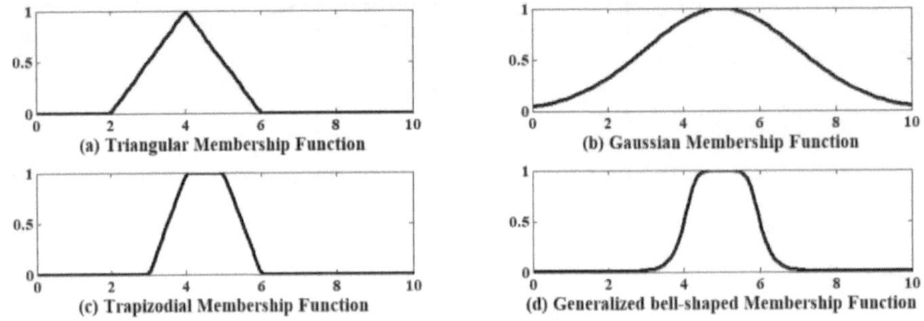

Figure 3.4 Four different types of fuzzy membership functions; (a) Triangular membership function, (b) Gaussian membership function, (c) Trapezoidal membership function, (d) Generalized bell-shaped membership function. (Matheworks, 2020.)

$$\mu_{\tilde{A}}(x;a,b,c,d) = \begin{cases} 0, & x \leq a. \\ \dfrac{x-a}{b-a}, & a \leq x \leq b. \\ 1, & b \leq x \leq c. \\ \dfrac{d-x}{d-c}, & c \leq x \leq d. \\ 0, & d \leq x. \end{cases} \tag{3.5}$$

The parameters $\{a,b,c,d\}$ (with $a < b \leq c < d$) represent the x coordinates of the four corners of the underlying trapezoidal membership function (a: lower boundary and d: upper boundary where membership degree is zero, b and c: the centers where membership degree is 1). Figure 3.4(c) illustrates the curve of the trapezoidal membership function defined by the parameters $a = 3$, $b = 4$, $c = 5$, and $d = 6$.

(iii) **Gaussian Membership Function** The Gaussian membership function is completely determined by two parameters $\{c,\sigma\}$ where $c,\sigma \in X$:

$$\mu_{\tilde{A}}(x;c,\sigma) = e^{-\frac{1}{2}\left(\frac{x-c}{\sigma}\right)^2} \tag{3.6}$$

Figure. 3.4(b) depicts the figure of the Gaussian membership function when $c = 5, \sigma = 5$.

(iv) **Generalized Bell-shaped Membership Function**

The generalized bell-shaped membership function is defined by three parameters $\{a,b,c\}$ where $a,b,c \in X$:

$$\mu_{\tilde{A}}(x;a,b,c) = \frac{1}{1+|\left(\frac{x-c}{a}\right)|^{2b}} \tag{3.7}$$

The parameter b is a positive value and Figure 3.4(d) plots the generalized bell membership function for the parameters $a = 1, b = 3, c = 5$.

(v) **The S-shaped Membership Function**(Medasani, Kim, and Krishnapuram, 1998)
The S-shaped membership function depends on two parameters a and b which locate the extremes of the sloped portion of the function, as given by Eq. 3.8.

$$\mu_{\tilde{A}}(x;a,b) = \begin{cases} 0, & x \leq a \\ 2\left(\frac{x-a}{b-a}\right)^2, & a \leq x \leq \frac{a+b}{2} \\ 1-2\left(\frac{x-b}{b-a}\right)^2, & \frac{a+b}{2} \leq x \leq b \\ 1, & b \leq x. \end{cases} \tag{3.8}$$

This function represents an asymmetrical polynomial curve open to the right and the graphical representation is depicted in Figure 3.5.

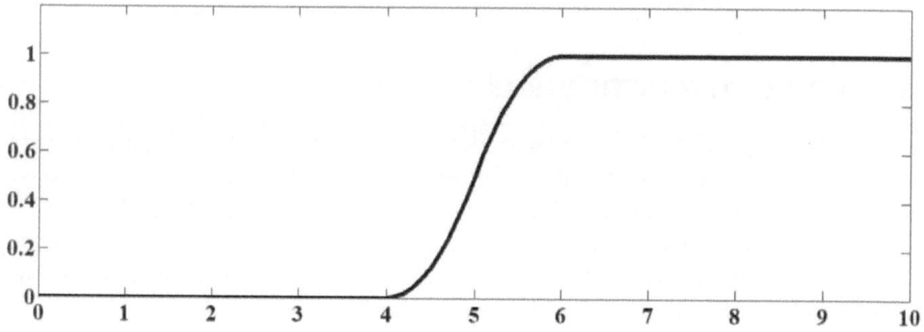

Figure 3.5 S-shaped membership function. (Matheworks, 2020.)

(vi) **The Z-shaped Membership Function**
The Z-shaped membership function is defined by two parameters a and b and represents an asymmetrical polynomial curve open to the left as defined in Eq. 3.9.

$$\mu_{\tilde{A}}(x;a,b)=\begin{cases}1, & x\leq a\\ 1-2\left(\dfrac{x-b}{b-a}\right)^2, & a\leq x\leq \dfrac{a+b}{2}\\ 2\left(\dfrac{x-a}{b-a}\right)^2, & \dfrac{a+b}{2}\leq x\leq b\\ 0, & b\leq x.\end{cases} \tag{3.9}$$

The figure of the Z-shaped membership function is depicted in Figure 3.6.

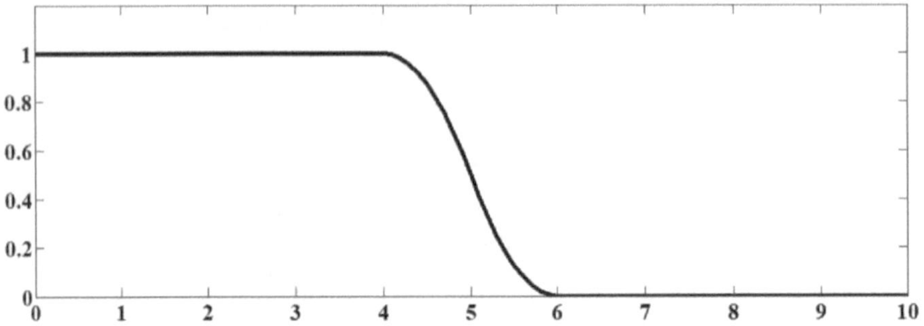

Figure 3.6 Z-shaped membership function. (Matheworks, 2020.)

3.3.3 ROLE OF LINGUISTIC VARIABLES

In most situations, when the universe of discourse X is in continuous space, usually the X is partitioned into several fuzzy subsets. The membership functions of those sets cover the set X relatively. Moreover, these sets are usually named linguistic labels or in a qualitative format such as "*High*," "*Medium*," and "*Low*." Thus the variable X takes the form of a linguistic/qualitative variable. The formal definitions and applications of linguistic variables can be found in later chapters.

Example 3.3 *Linguistic/Qualitative Variables*

Assume the universe of discourse X is Temperature. Then the fuzzy sets can be defined as "*Low*," "*Medium*," and "*High*" where the membership functions are defined as $\mu_{Low}(x)$, $\mu_{Medium}(x)$, $\mu_{High}(x)$ respectively.
The linguistic variable represents various values as an ordinary variable. For example, if "Temperature" takes the value of "*Low*" then one can express it as "Temperature is *Low*." The typical membership functions for these linguistic values are depicted in Figure 3.7.

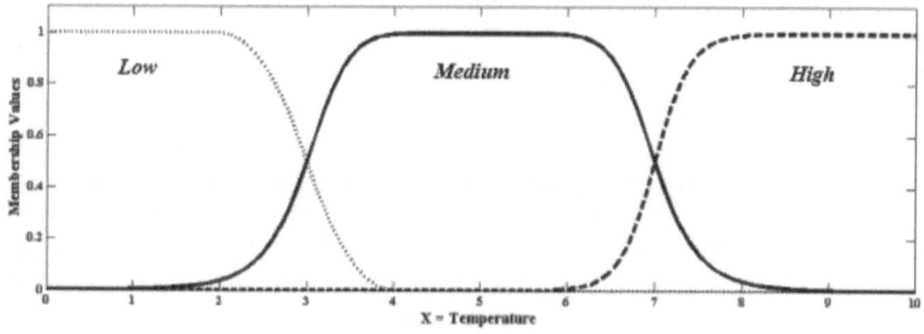

Figure 3.7 Membership functions of the linguistic values "*Low*," "*Medium*," and "*High*." (Matheworks, 2020.)

3.4 FUZZY SET - THEORETIC OPERATIONS

In classical set theory, union, intersection, and complement are the most basic operations. Fuzzy set theory has similar operations corresponding to the above-mentioned operations in classical set theory. First we present the definitions of these operations proposed by mathematician Zadeh in 1965. Since membership functions are the major component in fuzzy set theory, these operations are defined with the membership functions.

Definition 3.2. Intersection (Zimmermann, 2001)
The intersection of two fuzzy sets A and B is a fuzzy set C, represented by $C = A \cap B$ or $C = A$ AND B with corresponding membership functions of A and B by

$$\mu_{\tilde{C}}(x) = min(\mu_{\tilde{A}}(x), \mu_{\tilde{B}}(x)) = \mu_{\tilde{A}}(x) \wedge \mu_{\tilde{B}}(x). \quad (3.10)$$

The intersection of A and B is the largest fuzzy set which is contained in both A and B.

Definition 3.3. Union (Zimmermann, 2001)
The union of two fuzzy sets A and B is a fuzzy set C, represented by $C = A \cup B$ or $C = A$ OR B with corresponding membership functions of A and B by

$$\mu_{\tilde{C}}(x) = max(\mu_{\tilde{A}}(x), \mu_{\tilde{B}}(x)) = \mu_{\tilde{A}}(x) \vee \mu_{\tilde{B}}(x). \quad (3.11)$$

The union of A and B is the smallest fuzzy set containing both A and B.

Definition 3.4. Complement (Zimmermann, 2001)
The complement of a fuzzy set A, denoted by $\bar{A}$ ($\neg A$, **NOT**A), is defined as

$$\mu_{\bar{\tilde{A}}}(x) = 1 - \mu_{\tilde{A}}(x). \quad (3.12)$$

3.5 MORE ON FUZZY SET - THEORETIC OPERATORS

The fuzzy set theoretic operators may not be limited to the above-mentioned basic operators. There are many operators which have been suggested by mathematicians over the past four decades (Dubois and Prade, 1980b; Yager, 1980a, 235-242). According to the literature, there are two basic classes of operators: operators for intersection and union of fuzzy sets referred to as triangular norms (*t*-norms) and triangular co-norms (*t*-conorms) (Kóczy, 2014; Zhang, 2018; Liang, Zhao, and Luo, 2018; Jin, et al., 2019; Shi and Ye, 2018; Seikh and Mandal, 2019). The *t*-norms and *t*-conorms operators preserve monotonicity, commutativity, and associativity. Each class contains parameterized and non-parameterized operators. Table 3.1 presents some important non-parameterized operators. Now let us define some parameterized operators which are related to this study.

Table 3.1

Non-parameterized *t*-norms and *t*-conorms

t-norm	*t*-conorm
Minimum	**Maximum**
$\min\{\mu_{\tilde{A}}(x), \mu_{\tilde{B}}(x)\}$	$\max\{\mu_{\tilde{A}}(x), \mu_{\tilde{B}}(x)\}$
	Bounded sum
	$\min\{1, \mu_{\tilde{A}}(x)+\mu_{\tilde{B}}(x)\}$
Algebraic product	**Algebraic sum**
$\mu_{\tilde{A}}(x)\cdot\mu_{\tilde{B}}(x)$	$\mu_{\tilde{A}}(x)+\mu_{\tilde{B}}(x)-\mu_{\tilde{A}}(x)\cdot\mu_{\tilde{B}}(x)$
Hamacher product	**Hamacher sum**
$\dfrac{\mu_{\tilde{A}}(x)\cdot\mu_{\tilde{B}}(x)}{\mu_{\tilde{A}}(x)+\mu_{\tilde{B}}(x)-\mu_{\tilde{A}}(x)\cdot\mu_{\tilde{B}}(x)}$	$\dfrac{\mu_{\tilde{A}}(x)+\mu_{\tilde{B}}(x)-2\mu_{\tilde{A}}(x)\cdot\mu_{\tilde{B}}(x)}{1-\mu_{\tilde{A}}(x)\cdot\mu_{\tilde{B}}(x)}$
Einstein product	**Einstein sum**
$\dfrac{\mu_{\tilde{A}}(x)\cdot\mu_{\tilde{B}}(x)}{2-[\mu_{\tilde{A}}(x)+\mu_{\tilde{B}}(x)-\mu_{\tilde{A}}(x)\cdot\mu_{\tilde{B}}(x)]}$	$\dfrac{\mu_{\tilde{A}}(x)+\mu_{\tilde{B}}(x)}{1-\mu_{\tilde{A}}(x)\cdot\mu_{\tilde{B}}(x)}$

Definition 3.5. Hamacher (Zimmermann, 2001)
The Hamacher *t*-norm operator which is referred to as an "and" operator is defined as (intersection of two fuzzy sets $\tilde{A}$ and $\tilde{B}$):

$$\mu_{\tilde{A}\cap\tilde{B}}(x) = \frac{\mu_{\tilde{A}}(x)\,\mu_{\tilde{B}}(x)}{\gamma+(1-\gamma)\left(\mu_{\tilde{A}}(x)+\mu_{\tilde{B}}(x)-\mu_{\tilde{A}}(x)\,\mu_{\tilde{B}}(x)\right)}, \gamma \geqslant 0. \tag{3.13}$$

For $\gamma = 1$ the above operator reduces to an algebraic product.

The Hamacher *t*-conorm operator which is referred to as an "or" operator is defined as (union of two fuzzy sets $\tilde{A}$ and $\tilde{B}$):

$$\mu_{\tilde{A}\cup\tilde{B}}(x) = \frac{(\acute{\gamma}-1)\,\mu_{\tilde{A}}(x)\,\mu_{\tilde{B}}(x)+\mu_{\tilde{A}}(x)+\mu_{\tilde{B}}(x)}{1+\acute{\gamma}\mu_{\tilde{A}}(x)\,\mu_{\tilde{B}}(x)}, \acute{\gamma} \geq -1. \tag{3.14}$$

Definition 3.6. Yager (Zimmermann, 2001)
The Yager t-norm operator which is referred to as an "and" operator is defined as (intersection of two fuzzy sets $\tilde{A}$ and $\tilde{B}$):

$$\mu_{\tilde{A}\cap\tilde{B}}(x) = 1 - min\{1, ((1-\mu_{\tilde{A}}(x))^p + (1-\mu_{\tilde{B}}(x))^p)^{\frac{1}{p}}\}, p \geq 1. \quad (3.15)$$

The Yager t-conorm operator which is referred to as an "or" operator is defined as (union of two fuzzy sets $\tilde{A}$ and $\tilde{B}$):

$$\mu_{\tilde{A}\cup\tilde{B}}(x) = min\{1, (\mu_{\tilde{A}}(x)^p + \mu_{\tilde{B}}(x)^p)^{\frac{1}{p}}\}, p \geq 1. \quad (3.16)$$

For $p \to \infty$ the Yager t-norm operator converges to the min-operator and the t-conorm operator converges to the max-operator.

Definition 3.7. Dubois and Prade (Zimmermann, 2001)
The Dubois and Prade t-norm operator which is referred to as an "and" operator is defined as (intersection of two fuzzy sets $\tilde{A}$ and $\tilde{B}$):

$$\mu_{\tilde{A}\cap\tilde{B}}(x) = \frac{\mu_{\tilde{A}}(x)\cdot\mu_{\tilde{B}}(x)}{max\{\mu_{\tilde{A}}(x), \mu_{\tilde{B}}(x), p\}}, p \in [0,1]. \quad (3.17)$$

The t-conorm operator which is referred to as an "or" operator is defined as (union of two fuzzy sets $\tilde{A}$ and $\tilde{B}$):

$$\mu_{\tilde{A}\cup\tilde{B}}(x) = \frac{\mu_{\tilde{A}}(x) + \mu_{\tilde{B}}(x) - \mu_{\tilde{A}}(x)\cdot\mu_{\tilde{B}}(x) - min\{\mu_{\tilde{A}}, \mu_{\tilde{B}}, (1,p)\}}{max\{(1-\mu_{\tilde{A}}), (1-\mu_{\tilde{B}}), p\}}, p \in [0,1]. \quad (3.18)$$

The t-norm of Dubois and Prade lies between the min operation for $\alpha = 0$ and the algebraic product for $\alpha = 1$. And the t-conorm lies between the max operator for $\alpha = 0$ and the algebraic sum for $\alpha = 1$.
The last operator "Dombi" may be defined as follows:

Definition 3.8. Dombi (Seikh and Mandal, 2019)
The Dombi t-norm operator which is referred to as an "and" operator is defined as (intersection of two fuzzy sets $\tilde{A}$ and $\tilde{B}$):

$$\frac{1}{1 + \left[\left(\frac{1}{\mu_{\tilde{A}}} - 1\right)^p + \left(\frac{1}{\mu_{\tilde{B}}} - 1\right)^p\right]^{\frac{1}{p}}}, p \geqslant 1. \quad (3.19)$$

The t-conorm operator which is referred to as an "or" operator is defined as (union of two fuzzy sets $\tilde{A}$ and $\tilde{B}$):

$$\frac{1}{1 + \left[\left(\frac{1}{\mu_{\tilde{A}}} - 1\right)^{-p} + \left(\frac{1}{\mu_{\tilde{B}}} - 1\right)^{-p}\right]^{\frac{1}{p}}}, p \geqslant 1. \quad (3.20)$$

4 Modeling Quantitative Parameters

4.1 THE ROLE OF FUZZY SET THEORY IN MODELING QUANTITATIVE PARAMETERS

Quantitative parameters is a type of variable whose data is measured numerically and continuous in nature. When making judgments via quantitative parameters, one may somehow feel it is easy to handle those. But in some situations, it can be seen that uncertainty and imprecision arise in handling such parameters even though these parameters deal with numerical values. For example, assume one may need to find the tallest students in a particular classroom. Here the label "tallest" is determined by the height of a student. Now the question is, "What is the rigid boundary of the set of tallest students?" One can set the boundary as 167cm, another one can set 167.5cm, and so forth. For this example, the crisp set and a fuzzy set for tallest students have been set as depicted in Figure 4.1. The crisp set called "tallest students" is determined by the characteristic function where the boundary is 167cm, and the fuzzy set called "tallest students" is determined by the membership function which has no rigid boundary.

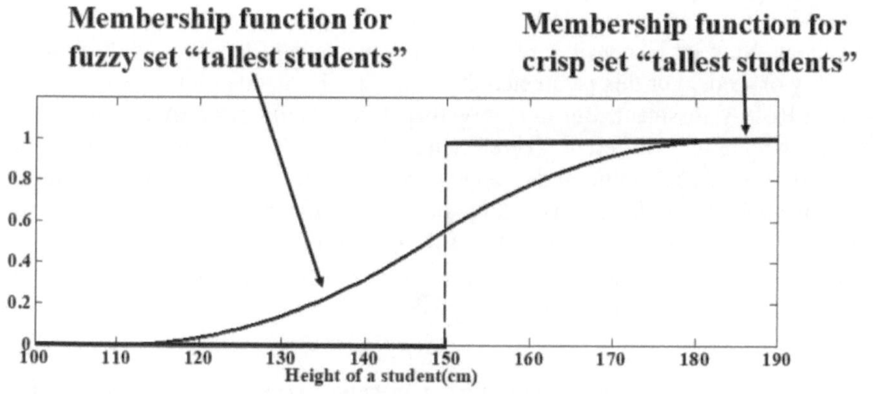

Figure 4.1 Crisp and fuzzy membership functions defined for the set *tallest students*. (Mathworks, 2020.)

Suppose one may find the tallest student set among the students s_1, s_2, s_3, s_4,s_5 using the characteristic function and membership function as shown in Table 4.1.

According to Table 4.1, it can be seen that the fuzzy membership function transformed the height into a degree of membership function such as 120cm into degree of membership value 0.044 out of 1, height 185cm into degree of membership value 0.73 out of 1, and so forth. One may note that the student with height 162cm is not in the crisp set whereas the student with height 169cm is in the set. Since the crisp set is accompanied with a rigid boundary the error occurring in the inclusion of an element would be high. But in the fuzzy set, the error occurring in the inclusion process spread over the set uniformly.

Table 4.1

Student's height in crisp and fuzzy form

Label	Height (cm)	Crisp value	Fuzzy value
s_1	120	0	0.044
s_2	185	1	0.73
s_3	162	0	0.43
s_4	169	1	0.53
s_5	172	1	0.57

Now let us discuss an example related to assessing the risk of an Invasive Alien Plant Species (IAPS).

Consider the parameter viability of seeds in months. First let us see how this parameter is evaluated in RA as in Figure 4.2. (Ranwala, 2010). According to Figure 4.2, one may note that any plant which undergoes the RA process obtains the score 1,2, and 3 if the viability is less than 3 years, 3-5 years, greater than 5 years, respectively. Now let us consider two plant species having 06 years and 20 years of viability of seeds. For this parameter, these two species get the same score which is 4 from Risk Assessment. But in reality, their level of impact on invasiveness with respect to viability is different. Therefore it is very important to obtain the degree of compatibility of each value in the range of a particular parameter. Thus parameter quantification or fuzzification is an important task to understand the behavior of the particular parameter which highly affects the decision making process.

In the next section, key points to develop fuzzy sets and membership functions for the quantitative parameters are presented.

4.2 FUZZY SETS AND MEMBERSHIP FUNCTIONS FOR QUANTITATIVE PARAMETERS

Fuzzification of parameters can be explained as the process of defining a fuzzy set and a suitable membership function for a particular parameter. This quantification process would not be an easy task if the given quantitative parameter were accompanied with ill-defined domains or no sharp boundaries (see Figure 4.1). Therefore in this section we propose a method to make the fuzzification process easier. This task mainly consists of two steps such as defining the fuzzy set and its membership

The three predefined answers in RA	Predefined scores provided in RA
i. Seeds remain viable up to 3 years	**1**
ii. Seeds remain viable up to 3-5 years	**2**
iii. Seeds remain viable > 5 years	**4**

Figure 4.2 Sample format of NRA. (Adapted from Ranwala, 2010.)

function.
Let us consider the first step:

1. Defining a fuzzy set for a particular quantitative parameter.

Assume we have been given a quantitative parameter with no rigid boundaries for fuzzification. In the fuzzification process, we need to set the fuzzy set. For that purpose, determining the lower limit and upper limit of the set is the vital part. These boundary points can be determined by the behavior of the given parameter by examining data. For such a judgment, the opinions of the experts in the relevant field can be considered, if given data is not sufficient. Also we have set up some assumptions which one may need to consider in determining the boundary points. These assumptions are:

- the lower boundary point of the particular fuzzy set is the lowest possible value which shows the minimum effect accompanied in the corresponding parameter.
- the upper boundary point of the particular fuzzy set is the extreme value which shows the maximum effect accompanied in the corresponding parameter.

Below we elaborate upon an example to show how to define a fuzzy set for a quantitative parameter based on the nature of the problem together with the available information.

Example 4.2.1

Here we consider the parameter viability of seeds in months defined as *VIA* which is one of the main factors in determining the dispersal risk of an IAPS.
First of all, one may need to have a clear idea of the objective of study to define the particular fuzzy set. According to this example, the objective is to quantify/fuzzify

this parameter in order to reflect the invasion risk accompanied with that parameter towards invasiveness. As we have mentioned earlier, the first one may be to define the domain, i.e. to determine the lower and upper boundary points. In this task, we have studied some risk assessments in several countries including Sri Lanka. Also the experts, opinions have also been taken to determine the boundary points. Therefore, the lower and the upper boundary points of the parameter *VIA* are 3 months and 1200 months respectively. The specified fuzzy set for the parameter *VIA* may now be defined as in Eq. (4.1):

$$\tilde{C} = \{(x, \mu_{VIA}(x)) \,|x\varepsilon R, \mu_{VIA}(x)\, \varepsilon \,[0,1]\}. \tag{4.1}$$

where $\mu_{VIA}(x)$ is the membership function. Now let us move to the second step.

2. Setting up the membership function.
Here we are concerned with setting the membership function of the particular fuzzy set in accordance with the objective of the study. There is no well-defined way to set up a membership function, and it may appear in any shape. Also several functions can be defined for a particular set. If so, one has to choose the most appropriate function. In most cases, the ultimate membership function which appears in the model may be the modified version of the function which was chosen initially. As for now, just the selection process of the appropriate membership function may be sufficient; other information can be found in the following sections.
Below we discuss the procedure to set up the appropriate membership function to its corresponding fuzzy set in detail by taking into consideration Example 4.2.1. In step 1, the fuzzy set for the parameter VIA has already been defined. Therefore, in this step, the membership function of set $\tilde{C}$, i.e. $\mu_{VIA}(x)$, needs to be defined. If one may look into the behavior of the parameter *VIA*, the risk towards invasiveness may increase after 3 months, and also it gives maximum effect after reaching 1200 months. It is not seen that any points between 3 to 1200 months act differently. Since these kinds of natural phenomena do not usually retain linear relationships, we have concluded that the most suitable function be a Z-shaped or S-shaped membership function (see Figures 3.5 and 3.6). The Z-shaped membership function for the set $\tilde{C}$ is as defined in Eq. (4.2):

$$\mu_{VIA}(x) = \begin{cases} 1 & \text{for } x \leq 3 \\ 1 - 2\left[(x-3)/1197\right]^2 & \text{for } 3 \leq x \leq \dfrac{1203}{2} \\ 2\left[(1200-x)/1197\right]^2 & \text{for } \dfrac{1203}{2} < x \leq 1200 \\ 0 & \text{for } x \geq 1200. \end{cases} \tag{4.2}$$

One may see that the membership function value of the fuzzy parameter $\tilde{C}$ is getting closer to zero when the seeds have a long viable period. In other words, if the viability is closer to 1200, the parameter value gives a value which is much closer to zero.

In this section we have discussed the parameter fuzzification process. In the next section, we discuss how to develop models using these fuzzy parameters and other related aspects.

4.3 MODEL FORMULATION

In the previous section, the fuzzification process of the quantitative parameters has been discussed. Here we propose a method to build up the required model using those fuzzified parameters.

To understand this method more clearly, let us discuss an example of assessing the dispersal risk of IAPS. As to literature, there are several influential risk factors which affect the dispersal risk of IAPS. Assume there are four quantitative parameters/factors which are highly dependent on dispersal and are quantified/fuzzified as was done in Section 4.2. As one may see what remains is to evaluate the combined effect of the four fuzzified parameters which reflect the dispersal risk of particular invasive plant species. In order to produce the combined effect of the parameters, the fuzzy sets with respect to these parameters have to be aggregated.

As we discussed in Chapter 2, non-parameterized and parameterized fuzzy operators play a vital role in aggregating fuzzy parameters. As such in the next subsections, we highlight the importance of aggregating fuzzy sets and the important aspects that have been followed to build up the required model.

4.3.1 AGGREGATION OF FUZZY SETS

Fuzzy set theory operators may be used to aggregate fuzzy sets/fuzzified parameters in the context of decision making. Certain properties such as handling cumulative effects, compensations, and interactions should be preserved along with the chosen operator to produce reliable output.

The importance of introducing such properties may be explained using an example given below:

One may assume seeds/per fruit, annual seed rain, viability of seeds, and long distance dispersal strength are the four quantitative parameters in identifying dispersal risk of IAPS.

It is not realistic to assume that species with extremely high viability of seeds have the potential to become more invasive. Apparently, the effect of high viability of seeds can be either amplified by the presence of an increase of annual seed rain and moderate level of long distance dispersal strength or compensated by low annual seed rain and moderate level of long distance dispersal strength. As mentioned in the literature, operators such as Hamacher, Yager, Dombi, and Dubois etc. are t-norm/t-conorm parameterized operators which preserve the properties mentioned above and have been widely applied in many areas (Lemaire, 1990).

4.3.2 IMPORTANT DEGREES OF THE PARAMETERS

Let us assume a set of parameters is given to assess a certain problem where the level of contribution of each parameter to the relevant problem is different from parameter to parameter, though it is not reasonable to assume that each parameter has the same importance for the specific problem. For example, we assumed that the viability of seeds is the better predictor of dispersal risk, while the importance of other parameters should be downgraded. Therefore, in these kinds of uncertain and imprecise situations, it is urged to weight those parameters.

There are several methods in the literature to weight the parameters (Zadeh, 1972). Among them, we reproduce two methods which are found to be more important. The first method is based on concentration/dilation operators which are directly used to upgrade or downgrade the parameters (Huynh, Ho, and Nakamori, 2002; Lemaire, 1990). The second method is finding weights using Chang's extent analysis and Buckly's methods by incorporating the experts, opinions. In the next subsection, we present more information regarding concentration/dilation operators.

4.3.3 CONCENTRATION/DILATION OPERATORS

Below we reproduced the concentration and dilation operators as in Lemaire (1990):

Definition 4.1. (Lemaire, 1990): The concentration of a fuzzy set $\tilde{A}$ is denoted by CON($\tilde{A}$) and may be expressed as

$$\mu_{CON(\tilde{A})}(x) = \mu_{\tilde{A}}^{\alpha}(x) \ \alpha > 1.$$

The operator concentration reduces the grade of membership of all the elements of a fuzzy set that are only partly in the set.

For example, if one may need to downgrade the value of μ_{VIA} when x is 48 months by 2 units then the corresponding downgraded value is $\mu_{CON(VIA)}(48) = \mu_{VIA}^{2}(48) = 0.9983^2 = 0.9966$.

Definition 4.2. (Lemaire, 1990): The dilation of a fuzzy set $\tilde{A}$ is denoted by DIL($\tilde{A}$) and may be expressed as

$$\mu_{DIL(\tilde{A})}(x) = \mu_{\tilde{A}}^{\alpha}(x) \ \alpha < 1.$$

Dilation is the opposite of concentration. A fuzzy set is dilated by increasing the grade of membership of all elements that are partly in the set.

For example, if one may need to upgrade the value of μ_{VIA} when x is 20 years (240 months) by 0.5 units then the corresponding upgraded value is $\mu_{DIL(VIA)}(240) = \mu_{VIA}^{0.5}(240) = 0.9527^{0.5} = 0.9983$. Figure 4.3 represents the graphical changes of the behavior of a membership function after applying CON, DIL operators.

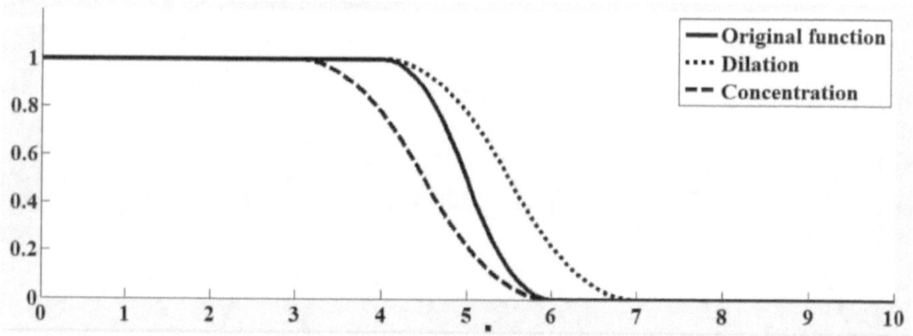

Figure 4.3 Concentration and dilation operators. (Matheworks, 2020.)

4.3.4 MODEL OUTPUT

So far, we have considered fuzzification and aggregation of quantitative parameters in an uncertain, imprecise environment. The decision, i.e. what we obtain as the output of the model, occurs from the aggregation process. Now the question is, "What is the appropriate way to interpret the output since those values are in the form of fuzzy values?" One way is to defuzzify the fuzzy value into a numerical/point value; however, the final output would be more informative if it were in the form of a range of values such as a linguistic term rather than a point/single-value. For example, in risk hazards, the final outcome usually interprets the severity of the risk such as in risk level: "*High*", "*Medium*", "*Low*".

By converting point-value into risk levels/linguistic terms, one can omit insignificant quantitative differences and focus more on qualitative study. For example, assume if one may model the dispersal risk of IAPS which gives risk levels as the output. And those levels may be categorized into seven risk levels such as Unlikely (U), Very Low (VL), Low (L), Medium (M), High (H), Very High (VH), Extremely High (EH). Figure 4.4 depicts the above-mentioned risk levels. The semantics of those labels are defined with fuzzy sets in order to capture the uncertainty of those linguistic terms.

Now let us see how the numerical value/fuzzy value may be obtained from aggregation process transforms into a risk level using fuzzy numerical-linguistic transformation functions. The numerical-linguistic transformation function for a given numerical value can be represented as follows (Delgado, et al., 1998):

Definition 4.3. Let $q\varepsilon\,[0,1]$ be a numerical value and l_i be a label verifying that $h(q,l_i)$=min$\{h(q,l_t)\,|\forall l_i\varepsilon L\}$, with

$$h(q,l_t)=\begin{cases} c & \text{if } q\notin \text{Supp }(l_t), \\ \Sigma_{j=1}^{c}\,(q-G_j(l_t))^2 & \text{if } q\in \text{Supp }(l_t). \end{cases} \tag{4.3}$$

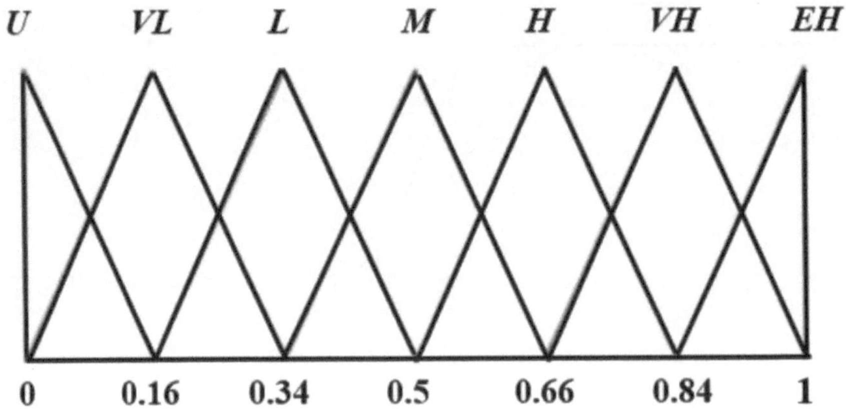

Figure 4.4 Membership functions of linguistic term set S. (Reprinted from H.O.W. Peiris, S. Chakraverty, S.S.N. Perera, and S.M.W. Ranwala. "Novel fuzzy linguistic based mathematical model to assess risk of invasive alien plant species," *Applied Soft Computing* 59, (2017): 326-339, Copyright (2017), with permission from Elsevier.)

where c is the cardinality of the characteristic function set G_j and Supp () refers to the support of a given membership function. The characteristic function set G_j consists of three functions which generate crisp values summarizing the information of a given fuzzy number (Delgado, Vila, and Voxman, 1998). In this task, three characteristic functions have been used by setting the cardinality, $c = 3$ to evaluate the label l_i. The three characteristic functions G_j are given in the formula Eqs.(3.10) to (3.14).

1. $G_1(l_i)$ - It is the method of center of gravity. This method summarizes the meaning of a label l_i into a numerical value as:

$$G_1(l_i) = \frac{\int_V v\mu_{y_{l_i}}(v)\,dv}{\int_V \mu_{y_{l_i}}(v)\,dv} \tag{4.4}$$

For the triangular fuzzy number, the function $G_1(l_i)$ may be expressed as:

$$G_1(l_i) = \begin{cases} x_1, & \text{if } x_1 = x_2 = x_3 \\ \dfrac{x_3^2 - x_1^2 + x_3x_2 - x_1x_2}{3(x_3 - x_1)} & \text{otherwise} \end{cases} \tag{4.5}$$

2. $G_1(l_i)$- This is the method of value of a fuzzy number:

$$G_2(l_i) = \int_0^1 s(r)\left(L_{y_{s_i}}(r) + R_{y_{s_i}}(r)\right)dr \tag{4.6}$$

where $L_y(r), R_y(r)$ are the r-cut representations of y_{s_i} and $s(r)$ is a reducing function.
The simplified form may be expressed by using the triangular fuzzy membership function and taking $s(r) = r$ as:

$$G_2(l_i) = x_2 + [(x_3 - x_2) - (x_2 - x_1)]/6 \tag{4.7}$$

3. $G_3(l_i)$-which is the method of *maximum value.*
Let us consider the given label as l_i, with a membership function $\mu_{y_{s_i}} = (v)$, $v \in V = [0,1]$. The height is defined as:
height(l_i)=Sup$\{\mu_{y_{s_i}}(v), \forall\mu\}$.

$$G_3(l_i) = \max\{v|\mu_{y_{s_i}}(v) = \text{height } (l_i) \tag{4.8}$$

For example, assume the value of q is 0.62; then the representative linguistic label is "*High*" where
$\min\{h(0.62,U), h(0.62,VL), h(0.62,L), h(0.62,M), h(0.62,H), h(0.62,VH), h(0.62,EH)\}$
$= \min\{1.0567, 0.6257, 0.2409, 0.0432, 0.0057, 0.1409, 0.376\} = 0.0057$
$=h(0.62,H)$

4.4 STRUCTURE OF THE MODEL–QUANTITATIVE PARAMETERS

Here we present the structure of a model which can be used in real applications. This structure is appropriate in decision making processes such as in risk assessments, factor-based problems where the related parameters are quantitative,

By considering facts discussed in the previous sections, the structure of the model has been designed as depicted as in Figure 4.5.
Let us now discuss the structure step by step. The first step is to collect the data for the selected parameters. Then define the fuzzy set and membership function for each of the selected parameters, i.e. the fuzzification process. Subsequently, an appropriate aggregation operator has to be selected. In that process, importance weights can be assigned if necessary. Here the complement of the final aggregated value has to be evaluated since we have chosen a $Z-$ shaped membership function. Now comes to the vital part, the validation process. If the complement of value obtained in the aggregation process illustrates the real scenario then the decision maker can terminate the process; otherwise the process is redirected to the fuzzification step. In that step, the membership functions should be modified and the suitable importance weights need to be incorporated with the fuzzified parameters in order to eliminate the deviations occurred in the validation step. These steps should run back and forth until the model generates a significant output.

The following section presents the procedures to evaluate the importance weights of the parameters using the experts' opinions and developing steps of models by incorporating those weights.

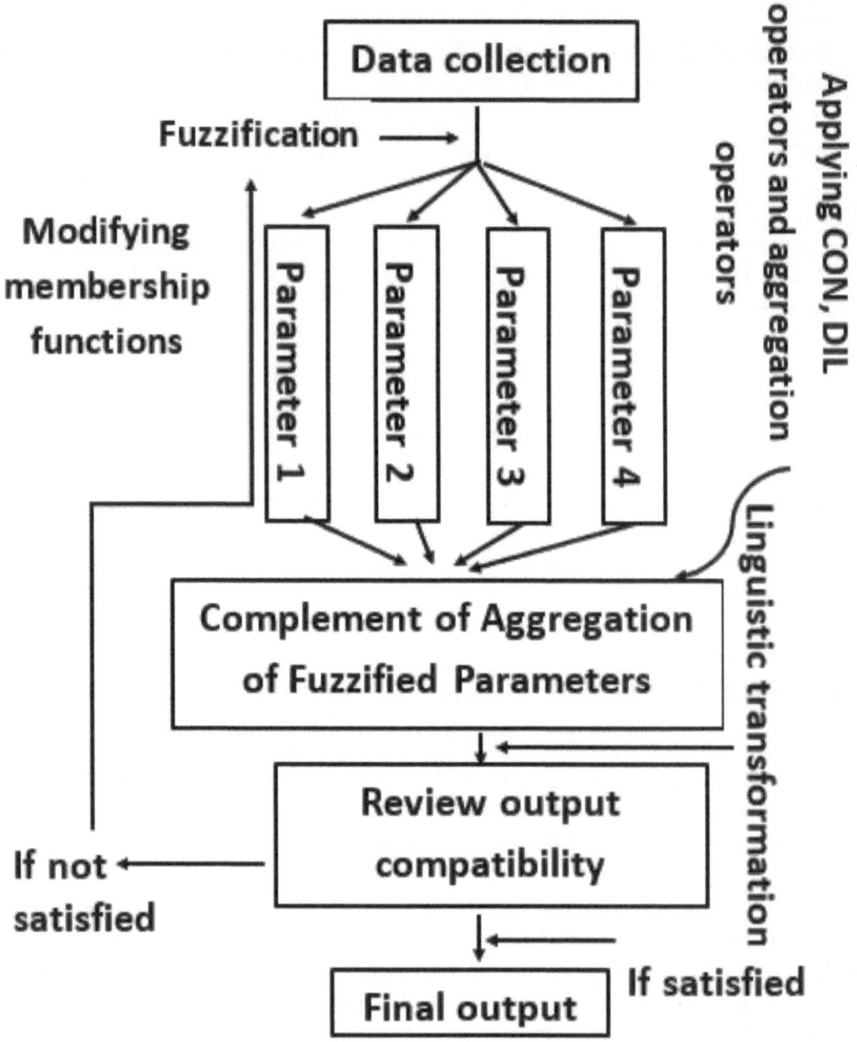

Figure 4.5 Model Structure - Quantitative parameters. Adapted from H.O.W. Peiris, S. Chakraverty, S.S.N. Perera, and S.M.W. Ranwala, "Modelling Dispersal Risk of Invasive Alien Plant Species," In *Recent Advances in Applications of Computational and Fuzzy Mathematics*. (Springer 2018) 109-145.

4.5 FUZZY ANALYTICAL HIERARCHY TECHNIQUES FOR EVALUATING GRADE OF IMPORTANCE WEIGHTS

4.5.1 AN OVERVIEW

In Section 4.3.2, one may see that the importance degrees of some parameters have been changed using CON, DIL operators. There are specific types of problems about which those in this field of expertise can give their judgments on the importance of the parameters using their knowledge and experience. This procedure is one of the knowledge elicitation methods which can be applied in an uncertain and imprecise situation. Usually experts may not use precise numbers to give preference information due to the complexity of the given problem. The use of linguistic terms/words instead of exact numbers provides decision makers the ability to make their judgment in a precise way. Sometimes the experts/decision makers find it difficult to express their judgments directly on a particular parameter. This may lead to providing their judgments by pairwise comparison of parameters. The techniques of the Fuzzy Analytical Hierarchy Process (FAHP) provide opportunities to handle such uncertain situations (Cinar, 2009; Saaty, 2008; Huang and Ho, 2013; Vaidya and Kumar, 2006). Here we elaborate upon two methods in FAHP, namely Chang's extent analysis and Buckly's method (Column geometric mean method) to evaluate the important degrees of given parameters. Further, developing steps of the factor-based models integrated with importance weights are being presented. In the next section we present some preliminaries which are relevant to this work.

4.5.2 PRELIMINARIES - EVALUATING THE GRADE OF IMPORTANCE WEIGHTS

The steps for evaluating the grade of importance weights using the FAHP technique incorporated with Chang's and Buckly's methods are as follows (Cinar, 2009; Huang and Ho, 2013):

Step 1: Evaluate the fuzzy reciprocal matrix

$$P = [\tilde{q}_{ij}], \tag{4.9}$$

where $\tilde{q}_{ij} = (l_{ij}, m_{ij}, u_{ij})$, l_{ij}, m_{ij}, and u_{ij} are the lower limit, peak, and upper limit of the triangular fuzzy number. $\tilde{q}_{ij} = \frac{1}{\tilde{q}_{ji}} = (\frac{1}{u_{ij}}, \frac{1}{m_{ij}}, \frac{1}{l_{ij}}), \forall i, j = 1, 2, ..., n$.

Step 2: Aggregate the experts' responses using the geometric mean method.

$$\tilde{q}_{ij} = (\tilde{q}^1_{ij} \bigotimes \tilde{q}^2_{ij} \bigotimes \cdots \bigotimes \tilde{q}^n_{ij})^{\frac{1}{n}}, \tag{4.10}$$

where $\tilde{q}_{ij}$ is the triangular fuzzy number in the ith column and j^{th} row of the fuzzy positive reciprocal matrix and $\tilde{q}^n_{ij}$ is the response value of the nth expert.

Step 3. Calculate the fuzzy importance weights.

Method I: Chang's extent analysis.

$$S_i = \sum_{j=1}^{m} M_{gi}^{j} \bigotimes \left[\sum_{i=1}^{n} \sum_{j=1}^{m} M_{gi}^{j} \right]^{-1}, \tag{4.11}$$

$$\sum_{j=1}^{m} M_{gi}^{j} = \left[\sum_{j=1}^{m} l_j, \sum_{j=1}^{m} m_j, \sum_{j=1}^{m} u_j \right], \tag{4.12}$$

$$\left[\sum_{i=1}^{n} \sum_{j=1}^{m} M_{gi}^{j} \right]^{-1} = \left[\frac{1}{\sum_{i=1}^{n} u_i}, \frac{1}{\sum_{i=1}^{n} m_i}, \frac{1}{\sum_{i=1}^{n} l_i} \right], \tag{4.13}$$

where S_i is the i_{th} fuzzy weight, in matrix m and $M_{gi}^{j}(j = 1, 2, ..., m)$ is the triangular fuzzy number calculated after comparing the questionnaires. After comparing each parameter, a minimum has been generated for each group as in Eq. (4.7).

$$V(M \geq M_1, M_2, ..., M_k) = minV(M \geq M_i), i = 1, 2, ..., k. \tag{4.14}$$

Assume that $d(A_i) = minV(S_t \geq S_k)$ for $k = 1, 2, ..., n; k \neq i$. Then the weight vector is given by

$$\acute{W} = (\acute{d}(A_1), (\acute{d}(A_2), ..., (\acute{d}(A_n))^T, \tag{4.15}$$

where $A_i, i = 1, 2, ..., n$.
Via normalization, the normalized weight vectors are

$$W = (d(A_1), (d(A_2), ..., (d(A_n))^T. \tag{4.16}$$

Method II: Buckly's method (Column geometric mean method).

$$\tilde{w}_i = r_i \bigotimes (r_1 \bigoplus r_2 \bigoplus ... \bigoplus r_n)^{-1}, \tag{4.17}$$

$$r_i = (\tilde{q_{ij}^1} \bigotimes \tilde{q_{ij}^2} \bigotimes ... \bigotimes \tilde{q_{ij}^n})^{\frac{1}{n}}, \tag{4.18}$$

where $\tilde{w}_i$ is the fuzzy weight value of each column in the fuzzy positive reciprocal matrix and r_i is the geometric mean of the triangular fuzzy number.

Step 4. Defuzzification of fuzzy weights into non-fuzzy values using Center of Gravity method (Bai and Wang, 2006).

4.5.3 CONSTRUCTION OF FUZZY PAIRWISE COMPARISON MATRIX

The fuzzy pairwise comparison matrix refers to the fuzzy reciprocal matrix as given in Eq.(4.9). The grade of importance weights of the parameters are derived from the pairwise comparison matrix where the entries of this matrix are the preferences of decision makers' judgments. The elements of the pairwise comparison matrices are presented as fuzzy numbers rather than exact numerical values in order to model the

Table 4.2

Linguistic scale with respective Triangular fuzzy number for Importance Adapted from H.O.W. Peiris, S. Chakraverty, S.S.N. Perera, and S.M.W. Ranwala, "Novel fuzzy based model on analysis of invasiveness due to dispersal related traits of plants," *Annals of Fuzzy Mathematics and Informatics*, 13, no(3) (2017): 6–14.

Linguistic Scale	Triangular Fuzzy Numbers
Absolutely more important	(5/2,3,7/2)
Very strongly more important	(2,5/2,3)
Strongly more important	(3/2,2,5/2)
Weakly more important	(1,3/2,2)
Equally important	(1/2,1,3/2)
Just equal	(1,1,1)

Parameters		Category 2			
		Parameter 1	Parameter 2	Parameter 3	Parameter 4
Category 1	Parameter 1	0	3		
	Parameter 2		0		
	Parameter 3			0	
	Parameter 4				0

Figure 4.6 Preference relation – I.

uncertainty and imprecision in the decision makers' judgments. In order to collect pairwise comparisons among the parameters, a questionnaire form needs to be constructed.

To fill out the questionnaire a panel of the field of experts has to be consulted to obtain the pairwise comparisons using an appropriate linguistic scale. Table 4.2 presents such a linguistic scale which can be used to collect the pairwise comparisons. Note that the cardinality of the terms set and its semantics can be changed by considering the nature of the problem and its parameters. As such, one can use trapezoidal fuzzy numbers instead of triangular fuzzy numbers.

The questionnaire has to be specially designed to perform all of the possible pairwise comparisons among risk factors. One may need to follow the scale to indicate the importance of one factor relative to another. A sample format of a questionnaire has been illustrated in Figures (4.6) and (4.7).

Figure 4.6 demonstrates the comparison among the parameters using the linguistic scale given in Table 4.2. (i.e. Category 1 $\longrightarrow$ 2). For instance, assume that the

Parameters		Category 2			
		Parameter 1	Parameter 2	Parameter 3	Parameter 4
Category 1	Parameter 1	0			
	Parameter 2	2	0		
	Parameter 3			0	
	Parameter 4				0

Figure 4.7 Preference relation – II.

importance of Parameter 1 compared to Parameter 2 is strongly more important toward invasiveness; then assign 3 in the corresponding cell in Figure 4.6. Assume that the importance of Parameter 2 compared to Parameter 1 is weakly more important; then assign 2 in the corresponding cell in Figure 4.7.

Following the above procedure one may be able to fill out the entries of the fuzzy pairwise comparison matrix.

4.5.4 MODEL FORMULATION

Proposed Method I

Here we propose a method to develop a model which consists of quantitative parameters with importance weights evaluated through Chang's extent analysis method or Buckly's method.
As the first step, one has to evaluate the grade of importance weights using pairwise comparisons as collected by the procedure mentioned in Subsection 4.5.3. Thereafter collected pairwise comparisons are integrated with the FAHP technique (Chang's method/Buckly's method) as described in Subsection 4.5.2.

The proposed method to construct a factor-based model by adopting the grades of importance weights which are evaluated from the Chang's method/Buckly's method is presented as follows:

Let X be the collection of objects denoted generically by x. Let $\mu_{P_1}(x)$, $\mu_{P_2}(x)$,...,$\mu_{P_n}(x)$ be the grades of membership of x with respect to parameters—Parameter 1, Parameter 2,.., Parameter n respectively. One may need to define the fuzzy set and its membership function for each parameter at the beginning of the process. Let us denote $w_i, i = 1, \dots, n$ as the normalized grade of importance weights obtained from either Chang's or Buckly's method for the Parameter 1, Parameter 2 to Parameter n respectively.
The proposed method for evaluating the effect of the n number of parameters in a particular scenario with respect to the object x:

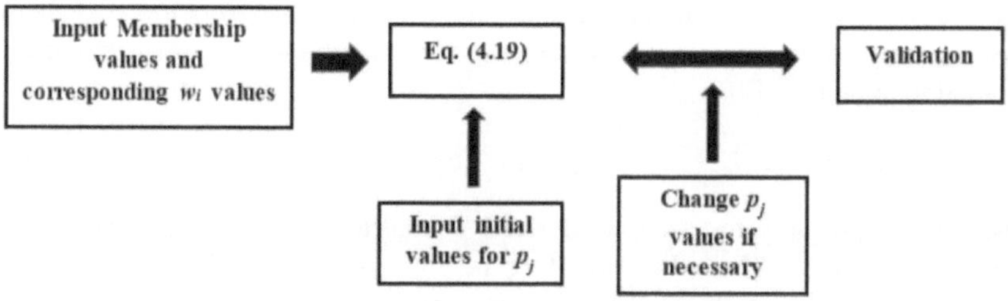

Figure 4.8 Determination of p_j values. (Adapted from H.O.W. Peiris, S. Chakraverty, S.S.N. Perera, and S.M.W. Ranwala, "Modelling Dispersal Risk of Invasive Alien Plant Species," In *Recent Advances in Applications of Computational and Fuzzy Mathematics*. (Springer 2018) 109-145.)

$$Output(x) = lin\left(w_1 \mu_{P_1}{}^{p_1} + w_2 \mu_{P_2}{}^{p_2} + \ldots + w_n \mu_{P_n}{}^{p_n}\right), x \in X, \tag{4.19}$$

where p_j ; j = 1,...,n are the unknown weights for the grade of membership of each factor. The symbol lin represents the numerical-linguistic transformation function where the final value in the form of the linguistic term is $Output(x)$.
The process of determining p_j values can be explained as shown Figure 4.8.

As depicted in Figure 4.8, at the beginning of the process, initial weights for p_j's should be defined and change until the model generates satisfactory output. Two different models can be obtained by applying the Chang's and Buckly's methods separately to the same set of data.
In the following subsection, we present another method which incorporates importance weights taken from two different methods.

Proposed Method II

In the previous subsection we have presented how to build up a model which is accompanied with importance weights. In that case, two methods have been illustrated to evaluate the importance weights. It is a notable fact that the existing weighting methods have both advantages and disadvantages. Therefore the preciseness of the weights may vary on those facts. One way to handle such scenarios is to combine the weights taken from different methods. Below we discuss the procedures to evaluate the combined weights and the model development.

Grade of Importance Weights

Here we assume that the importance weights are evaluated with two different methods. Then the importance weights are combined as a linear combination of weights derived from the two methods. The value λ has been incorporated as a characteristic value to adjust the weights derived from both methods to get the summation as the total weight as in Eq.(4.20).

$$W_T = \lambda W_I + (1-\lambda) W_{II} \tag{4.20}$$

where W_1 represents the weights derived from Method I, W_{II} represents the weights obtained from Method II, and λ is the characteristic value between 0 and 1. W_T is the final weight obtained for each parameter with respect to the value λ. Next we present the model which is accompanied with the new weights evaluated using the Eq. (4.20).

Model Development

Here, we assume that the model is in the form of an additive type model that consists of n parameters. The proposed model may be represented as follows:

$$Output(x) = lin\left(w_{1,\lambda}\mu_{P_1}^{p_1} + w_{2,\lambda}\mu_{P_2}^{p_2} + \ldots + w_{n,\lambda}\mu_{P_n}^{p_n}\right) \tag{4.21}$$

where p_j ; j = 1,...,n are the unknown weights for the grade of membership of each factor/parameter. The process of determining the p_j values may be executed similarly as in Figure 4.8.

As to Eq.(4.21) the values $w_{1,\lambda}$ to $w_{n,\lambda}$ with respect to λ are the grade of importance weights and μ_{P_1}, μ_{P_2},..., μ_{P_n} are the membership functions . The values p_1 to p_n are the weights that have to be determined for each membership value of the corresponding parameter. The symbol lin represents the numerical-linguistic transformation function where the final value in the form of the linguistic term is $Output(x)$.

In this chapter, we have presented factor-based models involved with quantitative parameters. Two different approaches to determine the importance weights of the parameters have also discussed.

5 Interval Function Approximation Techniques

5.1 AN OVERVIEW

In the situation where the problems involve uncertain and imprecise parameters, it is not reasonable to record data in point scale. In fact, considering the minimum and maximum recorded values offers a more complete insight about the phenomenon than considering the average values. The interval-valued data may contain imprecision and uncertainty for obtaining a reliable approximation. Therefore, many applications of computational research rely on interval function approximation. As mentioned in the literature (Hu, 2011), for many applications the explicit form of an interval function approximation is unavailable. To approximate a function, linear regression techniques may be used. One more advantage of linear regression is the ability to handle the quantitative and qualitative parameters concurrently (Stockburger, 1998).

There exist few works which have used complex computational methods to solve the interval least squares (Bentbib, 2002). In this work, three approaches have been considered to develop a factor-based model which consists of both quantitative and qualitative parameters. Approach I is based on the interval least squares algorithm proposed by Chenyi Hu (Hu, 2011). Approaches II and III use two newly implemented methods to estimate the interval-valued regression coefficients. In the next section, we present important preliminaries of interval linear regression that are used in this work.

5.2 PRELIMINARIES

5.2.1 BASIC CONCEPTS OF MULTIVARIATE LINEAR REGRESSION

Linear regression is one of the fundamental models used to determine the relationship between dependent and independent variables. An extension of this model, namely multiple linear regression, is used to represent the relationship between a dependent variable and several independent variables. These variables usually are single-valued. Here the interval-valued data x can be written by the pair of values $\underline{x}$ and $\overline{x}$ with $\overline{x} \geq \underline{x}$ where $\underline{x}$ and $\overline{x}$ denote the lower and upper bounds respectively.
Multivariate regression is an extension of simple linear regression in which more than one independent variable (x) is used to predict a single dependent variable (y) (Stockburger, 1998) .

The predicted value of y is a linear combination of the x variables such that the sum of squared deviations of the observed and predicted y is a minimum. The multiple linear regression equation for p variables is as follows:

$$y_i = a_0 + a_1 x_{i1} + a_2 x_{i2} + ... + +a_p x_{ip}, i = 1, 2, ..., n, \tag{5.1}$$

where y_i is the predicted or expected value of the dependent variable, x_{i1} through x_{ip} are p distinct independent or predictor variables, and a_0 through a_p are the regression coefficients to be estimated utilizing the input and output data. The regression coefficients $(a_i, i = 1, 2, ..., p)$ can be determined by using least squares estimation.

5.2.2 LEAST SQUARES ESTIMATION

This method can be used to estimate the regression coefficients in the linear regression model by minimizing the squared discrepancies between observed data and their expected values (Van De Geer, 2005). To find the least squares estimators, it will be more convenient to use the matrix notation of Eq. (5.1).

Let Y be a $p-$dimensional vector consisting of expected values, and X is an $n \times p$ data matrix of n observations of the p variables; then we have

$$X = \begin{pmatrix} 1 & x_{12} & . & . & . & x_{1p} \\ . & . & . & . & . & . \\ . & . & . & . & . & . \\ . & . & . & . & . & . \\ 1 & x_{n2} & . & . & . & x_{np} \end{pmatrix} \tag{5.2}$$

a is the $p-$dimensional vector of unknown regression coefficients. The least squares estimator, denoted by $\widehat{a}$ is that value of a that minimizes the

$$||Y - Xa||^2 \tag{5.3}$$

Suppose X has full column rank; that is no column in X can be written as a linear combination of other columns. Then the least squares estimator $\widehat{a}$ is given by

$$\widehat{a} = \left(X^T X\right)^{-1} X^T y. \tag{5.4}$$

Next we define the interval multiple linear regression model for interval input-output data.

5.2.3 INTERVAL MULTIVARIATE LINEAR REGRESSION MODEL

The Interval Multivariate Linear Regression model can be defined as follows:
The situation of a linear dependence of a variable $\tilde{Y}$ on m variables is of the form

$$\tilde{Y} = \theta_0 + \theta_1 \tilde{X}_{i1} + \theta_2 \tilde{X}_{i2} + \dots + \theta_m \tilde{X}_{im}, i = 1, 2, \dots, N. \tag{5.5}$$

Assume that the variables $\tilde{X}_{i1}$, $\tilde{X}_{i2}, \dots, \tilde{X}_{im}$, are given by interval observations $\left[\underline{X_{mi}}, \overline{X_{mi}}\right]$, $i = 1, 2, \dots, N > m$. Similarly the dependent variable $\tilde{Y}$ is considered as $\left[\underline{y}_i, \overline{y}_i\right]$, $i = 1, 2, \dots, N$. The interval vector $\theta = (\theta_0, \theta_1, \theta_2, \dots, \theta_m)^T$ represents the unknown regression coefficients.

5.2.4 INTERVAL LEAST SQUARES

The concept of the interval least squares can be applied to the interval valued linear system in Eq. (5.5) to obtain interval estimates of θ. In the present study, we basically concern on minimizing the overall absolute error of approximation.

Let us define the absolute error of interval estimation as in (Hu, 2011).

Definition 5.1. Let interval $\widehat{y} = \left[\widehat{\underline{y}}, \widehat{\overline{y}}\right]$ be an estimation of an interval $y = \left[\underline{y}, \overline{y}\right]$. The left and right absolute errors are $E_L = |\widehat{\underline{y}} - \underline{y}|$ and $E_R = |\widehat{\overline{y}} - \overline{y}|$, respectively. The absolute error of the estimation is the sum of left and right absolute errors, that is, $E = E_L + E_R = |\widehat{\underline{y}} - \underline{y}| + |\widehat{\overline{y}} - \overline{y}|$, respectively.

Using Definition 5.1, we define the sum of squares error (SSE) of the interval valued multiple linear regression system as in (Hu, 2011).

Definition 5.2. Let U be the set of n interval valued observations of an interval linear regression function $y = h(x)$, i.e.
$U = \{(x, y) : x \subset \Re^n, y \subset \Re, \text{both } x \text{ and } y \text{ are compact}\}$. We say that

$$\sum_{0 \le j \le m} \tilde{\theta}_j \tilde{X}_{ij}$$

is an interval least squares approximation of $h(x)$ if the linear combination minimizes:

$$\sum_{i=1}^{N} E_L^2 + \sum_{i=1}^{N} E_R^2,$$

where

$$\sum_{i=1}^{N} E_L^2 = \sum_{i=1}^{N} \left(\underline{y}_i - \left(\underline{\sum_{0 \le j \le m} \theta_j X_{ij}} \right) \right) \tag{5.6}$$

$$\sum_{i=1}^{N} E_R^2 = \sum_{i=1}^{N} \left(\overline{y}_i - \left(\overline{\sum_{0 \le j \le m} \theta_j X_{ij}} \right) \right) \tag{5.7}$$

5.2.5 INTERVAL ARITHMETIC

While using the Interval Least Squares (ILS) method, it is required to find the estimates for interval regression coefficients that minimize the sum of Eqs. (5.6) and (5.7). Therefore we need to use interval arithmetic to work with interval-valued data (Alefeld and Mayer, 2000).
Let $[a] = [\underline{a}, \overline{a}]$, and $[b] = [\underline{b}, \overline{b}]$ be real compact intervals and $\circ$ be one of the basic operations "addition", "subtraction", "multiplication", and "division", respectively (for real numbers) that is $\circ \in \{+, -, \cdot, \div\}$.

Then, $[a] \circ [b] = \{a \circ b | a \in [a], b \in [b]\}$. If $\circ$ is $\div$ then $0 \notin [b]$.

For the corresponding operations:

$$[a] + [b] = [\underline{a} + \underline{b}, \overline{a} + \overline{b}],$$

$$[a] - [b] = [\underline{a} - \overline{b}, \overline{a} - \underline{b}],$$

$$[a] \cdot [b] = [\min\{\underline{a} \cdot \underline{b}, \underline{a} \cdot \overline{b}, \overline{a} \cdot \underline{b}, \overline{a} \cdot \overline{b}\}, \max\{\underline{a} \cdot \underline{b}, \underline{a} \cdot \overline{b}, \overline{a} \cdot \underline{b}, \overline{a} \cdot \overline{b}\}],$$

$$[a] \div [b] = [\min\{\underline{a} \div \underline{b}, \underline{a} \div \overline{b}, \overline{a} \div \underline{b}, \overline{a} \div \overline{b}\}, \max\{\underline{a} \div \underline{b}, \underline{a} \div \overline{b}, \overline{a} \div \underline{b}, \overline{a} \div \overline{b}\}],$$
provided $0 \notin [b]$.

In the next section, we discuss the developing steps for the factor-based model.

5.3 FACTOR-BASED MODEL WITH INTERVAL INPUTS AND OUTPUTS

First we elaborate upon the assumptions made in the model development as follows:

- parameter may be in the form of quantitative or qualitative
- data for the given parameters are in the form of intervals
- assess the linear combination effect of the parameters
- model in the form of interval function, approximate via interval multivariate regression technique

Now let us define the model which follows the above assumptions via interval multivariate regression as

$$[\underline{y_i}, \overline{y_i}] = [\underline{a_0}, \overline{a_0}] + [\underline{a_1}, \overline{a_1}] [\underline{x_{i1}}, \overline{x_{i1}}] + \ldots + [\underline{a_m}, \overline{a_m}] [\underline{x_{ij}}, \overline{x_{ij}}] \quad \text{for} \quad i = 1, 2, \ldots, n \tag{5.8}$$

where $\tilde{y}_i$ is the interval response variable, $x_{\tilde{i}m}$, $j = 0, 1, \ldots, m$ are the interval predictor variables, and $\tilde{a}_j$'s are unknown interval regression coefficients. The task remains here to approximate the unknown coefficients $\tilde{a}_j$ to complete the model formulation process. Since there is no explicit method to find the inverse of an interval matrix which requires the interval least squares method one may need to move to approximation methods.

In the subsequent sections we now discuss Approaches I to III.

5.4 APPROACH I

In this approach, the regression coefficients are approximated based on the interval least squares algorithm proposed by Chenyi Hu (Hu, 2011). Let us now discuss the steps for determining the unknown interval regression coefficients using the algorithm mentioned above.

The process of finding coefficients may be executed with two steps. The first step is the initial solution and the second step is width adjustments for the initial point solution. These two steps are given in detail as mentioned in (Hu, 2011).

- **Initial Solution**
  Let $\tilde{X}\theta = \tilde{Y}$ be a simplified form of the interval linear equation Eq. (5.8). It is a computational challenge to solve $\tilde{X}^T\tilde{X}\theta = \tilde{Y}$ for θ directly or to perform the QR factorization on the interval-valued matrix. Therefore, this work mainly aims to find an approximated solution to meet our needs. First we find the midpoint solution for θ in the interval normal equation $X\theta = Y$ as the initial point solution. The algorithm for finding the midpoint solution is given in (Hu, 2011) as follows:
  For a given set of interval-valued pairs (x_i, y_i)

  - Evaluate the interval matrix $\tilde{X}$ as defined in Eq. (5.2).
  - Perform QR factorization on midpoint matrix, X_{mid} of $\tilde{X}$ such that $X_{mid} = QR$.
  - Calculate $c = Q^T\tilde{Y}$ with interval arithmetic.
  - Compute midpoint solution of θ by solving $R\theta = c_{mid}$ with backward substitution.

- **Width Adjustments for Initial Point Solution**
  One may note that the initial point solution which is obtained from the above algorithm is a point solution. In order to convert the point solution to interval values, a sequence of $\varepsilon-$ inflations of $1\%, 2\%, \ldots, 10\%$ have been used.
- **Concept of Ratio Estimation**
  The concept of ratio estimation facilitates finding a satisfactory approximated solution. Applying the concept of volume of an interval vector $V = (q_1, q_2, \ldots, q_n)$, which is $v(q) = \prod_{1 \le i \le n} (\overline{q_i}, \underline{q_i})$, the notion of ratio of estimation for interval linear systems $X\theta = Y$ is reproduced as defined as in (Hu, De Korvin, and Kreinovich, 2008):

$$r = \begin{cases} 0\% & \text{if } X\theta \bigcap Y = \emptyset; \\ 100\% & \text{if } X\theta = Y; \\ \dfrac{\nu(Y)}{\nu(X\theta)} & \text{if } X\theta \supset Y; \\ \dfrac{\nu(X\theta)}{\nu(Y)} & \text{if } X\theta \subset Y; \\ \dfrac{\nu(X\theta \cap Y)}{\nu(X\theta \cup Y)} & \text{otherwise.} \end{cases} \tag{5.9}$$

In order to find the most suitable approximated solution, ratio estimation is calculated for each of the approximated solutions which may be obtained by various interval-valued θ. Among them, select the interval-valued θ with the highest ratio estimation which can be considered as the approximated interval regression coefficients of the model defined in Eq. (5.8).
Below we now propose a new method to find the estimations for interval regression coefficients.

5.5 APPROACH II

The problem in Approach I is adjusting widths of center values of regression coefficients using $\varepsilon-$ inflation of $1\%, 2\%, \ldots, 10\%$. But, if the center values of coefficients and boundaries of input and output data are positive we can directly find the boundary estimations for regression coefficients. In this section we propose a new method to estimate the interval regression coefficient of a multivariate regression model which satisfies certain conditions. Let us recall the interval multivariate regression model as defined in Eq.(5.8):

$$\left[\underline{y_i}, \overline{y_i}\right] = \left[\underline{a_0}, \overline{a_0}\right] + \left[\underline{a_1}, \overline{a_1}\right]\left[\underline{x_{i1}}, \overline{x_{i1}}\right] + \ldots + \left[\underline{a_j}, \overline{a_j}\right]\left[\underline{x_{ij}}, \overline{x_{ij}}\right] \text{for} i = 1, 2, \ldots, n \tag{5.10}$$

where $\tilde{y_i}$ is the interval response variable, $\tilde{x_{ij}}$, $j = 0, 1, \ldots, m$ are the interval predictor variables, and $\tilde{a_j}$'s are unknown interval regression coefficients. The center values of interval regression coefficients $\left[\underline{a_j}, \overline{a_j}\right]$ for $j = 0, 1, \ldots, m$ are considered as $\acute{a}_j$. We consider $\varepsilon_j > 0$ a value which satisfies:

$$\underline{a}_j = \acute{a}_j - \varepsilon_j, \tag{5.11}$$

$$\overline{a_j} = \acute{a}_j + \varepsilon_j, \text{for} j = 1, 2, \ldots, m. \tag{5.12}$$

If $\acute{a}_j > 0$ and $\underline{x_{ij}}, \overline{x_{ij}} \geq 0$ for each $j = 0, 1, \ldots, m$ and $i = 0, 1, \ldots, n$ then Eq.(5.10) can be written by using interval arithmetic mentioned in Section 5.2.5 as follows:

$$\underline{y_i} = (\acute{a}_0 - \varepsilon_0) + \min\left((\acute{a}_1 - \varepsilon_1)\underline{x_{i1}}, (\acute{a}_1 - \varepsilon_1)\overline{x_{i1}}\right) + \min\left((\acute{a}_2 - \varepsilon_2)\underline{x_{i2}}, (\acute{a}_2 - \varepsilon_2)\overline{x_{i2}}\right) + \ldots + \min\left((\acute{a}_m - \varepsilon_m)\underline{x_{in}}, (\acute{a}_m - \varepsilon_m)\overline{x_{in}}\right) \tag{5.13}$$

$$\overline{y_i} = (\acute{a}_0 + \varepsilon_0) + (\acute{a}_1 + \varepsilon_1)\overline{x_{i1}} + \ldots + (\acute{a}_m + \varepsilon_m)\overline{x_{im}}. \tag{5.14}$$

One may note that all $\underline{y_i}$,$\overline{y_i}$,$\underline{a_j}$,$\overline{a_j}$,$\underline{x_{ij}}$, and $\overline{x_{ij}}$'s are point values and not intervals. The most important observation here is that without knowing the sign of $\underline{a_j}$ it is not easy to find the estimations for $\underline{a_j}$ in Eq.(5.13). In order to find the estimates for $\underline{a_j}$ first we find estimates for $\overline{a_j}$.

Rearranging Eq.(5.14) we have

$$\overline{y_i} = \overline{c_0} + \overline{c_1 x_{i1}} + \ldots + \overline{c_m x_{im}}, \tag{5.15}$$

where $\overline{c_j} = \acute{a}_j + \varepsilon_j$ for $j = 1, 2, ..., m$.

Using the least squares method we find estimates for $\overline{c_j}$'s. Solving the equation $\overline{c_j} = \acute{a}_m + \varepsilon_j$ for ε_j we estimate the lower boundary estimation $\underline{a_j}$ as $\underline{a_j} = \acute{a}_j - \varepsilon_j$. Likewise we can estimate the interval regression coefficients of Eq.(5.10).

Accuracy Assessment

Here we discuss the procedure to examine whether the model produces satisfactory output with the approximated interval regression coefficients.
The quality of the approximation of $\tilde{Y}_i \approx \left(\sum_{0 \le j \le m} \tilde{\theta}_j \tilde{X}_{ij}\right)$ can be examined by considering the overlap between approximated output $\hat{\tilde{Y}}$ and expected output $\tilde{Y}$. The approximation would be better if the overlap between $\hat{\tilde{Y}}$ and $\tilde{Y}$ is considerably large.

Accuracy Ratio

The accuracy ratio of an interval approximation is defined in (Hu, 2011) as below.

Definition 5.3. Let $\hat{\tilde{Y}} = \left[\underline{\hat{y}}, \overline{\hat{y}}\right]$ be an approximation for the interval $\tilde{Y} = \left[\underline{y}, \overline{y}\right]$.The accuracy ratio of the approximation is

$$Acc\left(\tilde{Y}, \hat{\tilde{Y}}\right) = \begin{cases} 100\% & \text{if } \tilde{Y} = \hat{\tilde{Y}} \\ \dfrac{w\left(\left[\underline{y}, \overline{y}\right] \cap \left[\underline{\hat{y}}, \overline{\hat{y}}\right]\right)}{w\left(\left[\underline{y}, \overline{y}\right] \cup \left[\underline{\hat{y}}, \overline{\hat{y}}\right]\right)} & \text{if } \left(\tilde{Y} \cap \hat{\tilde{Y}}\right) \neq \emptyset \\ 0 & \text{otherwise} \end{cases} \tag{5.16}$$

where the function $w()$ returns the width of an interval.

Average Accuracy Ratio

For a set of N interval pairs (x_i, y_i) the average accuracy ratio of the approximation is defined as

$$Acc^* = \frac{\Sigma_{i=1}^{N} Acc\left(\tilde{Y}_i, \hat{\tilde{Y}}(x_i)\right)}{N}. \tag{5.17}$$

The average accuracy ratio is a quality measurement in addition to the sum of squares of left and right errors defined in Eqs. (5.6) and (5.7). Maximizing the average accuracy ratio and minimizing the sum of squares are inter-connected. The higher the average accuracy ratio is the smaller the sum of squares and the better the approximation.

5.6 APPROACH III

In this section, we are going to present another new estimation method to approximate the interval regression coefficients in an interval multiple linear regression model. In this task, we follow a different approach to develop the estimation method compared to the Approach II.
The steps for this estimation method are as follows:

First recall the interval multivariate linear regression model as stated in Eq.(5.8):

$$\left[\underline{y_i}, \overline{y_i}\right] = \left[\underline{a_0}, \overline{a_0}\right] + \left[\underline{a_1}, \overline{a_1}\right]\left[\underline{x_{i1}}, \overline{x_{i1}}\right] + ... + \left[\underline{a_j}, \overline{a_j}\right]\left[\underline{x_{ij}}, \overline{x_{ij}}\right] \text{ for} i = 1, 2, ..., n. \tag{5.18}$$

where $\tilde{y_i}$ is the interval response variable, $\tilde{x_{ij}}$, $j = 0, 1, ..., m$ are the interval predictor variables, and $\tilde{a}_j$'s are unknown interval regression coefficients.

Let us denote the center values of interval regression coefficients $\left[\underline{a_j}, \overline{a_j}\right]$ for $j = 0, 1, ..., m$ as $\acute{a}_{mid_j}$. To find $\acute{a}_m$, first of all we need to construct the matrix $\tilde{A}$ and vector $\tilde{b}$ as in Eqs. (5.2) and (5.3).

Let us take A_{mid} and b_{mid} as the midpoint matrices of $\tilde{A}$, $\tilde{b}$ respectively. One may note that all components in matrices A_{mid} and b_{mid} are crisp values, that is not intervals. If A_{mid} is a full column rank matrix we can evaluate $\acute{a}_{mid_j}$ as below:

$$\acute{a}_{mid_j} = \left(A_{\text{mid}}^T A_{\text{mid}}\right)^{-1} b_{\text{mid}}. \tag{5.19}$$

Now we consider $\varepsilon_j > 0$ a value which satisfies:

$$\underline{a_j} = \acute{a}_{mid_j} - \varepsilon_j, \tag{5.20}$$

$$\overline{a_j} = \acute{a}_{mid_j} + \varepsilon_j, \text{for } j = 1, 2, ..., m. \tag{5.21}$$

If boundary values of $\tilde{Y_i}$, $\tilde{x_{ij}}$ and center values $\acute{a}_{mid_j}$ are non-negative then the lower and upper observed responses satisfy the Eqs. (5.22) and (5.23) respectively.

$$\underline{y_i} = \underline{a_0} + \min\left(\underline{a_1 x_{i1}}, \underline{a_1 \overline{x_{i1}}}\right) + \min\left(\underline{a_2 x_{i2}}, \underline{a_2 \overline{x_{i2}}}\right) + \ldots + \min\left(\underline{a_n x_{in}}, \underline{a_n \overline{x_{in}}}\right)\ i = 1, 2, \ldots, n \tag{5.22}$$

$$\overline{y_i} = \overline{a_0} + \overline{a_1 x_{i1}} + \overline{a_2 x_{i2}} + \ldots + \overline{a_n x_{in}} i = 1, 2, \ldots, n \tag{5.23}$$

where $\underline{y_i}, \underline{x_{ij}}$ and $\underline{a_j}$ indicate lower bounds of response variables, predictor variables, and model parameters respectively. Similarly $\overline{y_i}, \overline{x_{ij}}$, and $\overline{a_j}$ indicate upper bounds of response variables, predictor variables, and regression coefficients respectively. The sign of $\underline{a_j}$ can be changed depending on the values of ε_j. Therefore without knowing the exact ε_j values, Eq. (5.22) cannot be computed as Eq. (5.23). Let us denote $[\hat{y_L}, \hat{y_U}]$ as the estimated interval output, where $\hat{y_L} = \left(\underline{a_0} + \sum_{1 \le j \le m} \min\left(\underline{a_j x_{ij}}, \underline{a_j \overline{x_{ij}}}\right)\right)$ and $\hat{y_U} = \left(\sum_{0 \le j \le m} \overline{a_j x_{ij}}\right)$.

Due to the complexity of finding the lower boundaries for regression coefficients in Eq.(5.22), it is urged to find a subset of $[\hat{y_L}, \hat{y_U}]$. Let us define $[\acute{y_L}, \hat{y_U}]$ as the subset of $[\hat{y_L}, \hat{y_U}]$ where $\acute{y_L} = \left(\underline{a_0} + \sum_{1 \le j \le m} \underline{a_j x_{ij}}\right)$. One may note that the defined subset differs only from the lower bound when it is compared with the set $[\hat{y_L}, \hat{y_U}]$.

Now let us show that $[\acute{y_L}, \hat{y_U}] \subseteq [\hat{y_L}, \hat{y_U}]$ if boundary values of $\tilde{Y_i}, \tilde{x_{ij}}$, and center values of $\acute{a}_{mid_j}$ are non-negative.

For the case when $\underline{a_j} \ge 0$, $\hat{y_L} = \acute{y_L} = \left(\underline{a_0} + \sum_{1 \le j \le m} \underline{a_j x_{ij}}\right)$ since $\underline{x_{ij}}$, $\overline{x_{ij}}$ are non-negative values.

Hence $[\acute{y_L}, \hat{y_U}] \subseteq [\hat{y_L}, \hat{y_U}]$.
Now consider the case when $\underline{a_j} < 0$. Without loss of generality we may assume that $\underline{a_1} < 0$ in Eq. (5.22).

Then it is clear that $\underline{a_1 \overline{x_{ij}}} < \underline{a_1 x_{ij}}$ since $\underline{x_{ij}}, \overline{x_{ij}}$ are non-negative values and $\underline{x_{ij}} < \overline{x_{ij}}$. By adding the remaining right hand terms for both sides of $\underline{a_1 \overline{x_{ij}}} < \underline{a_1 x_{ij}}$ we have

$$\underline{a_0} + \underline{a_1 \overline{x_{i1}}} + \underline{a_2 x_{i2}} + \ldots + \underline{a_n x_{in}} < \underline{a_0} + \underline{a_1 x_{i1}} + \underline{a_2 x_{i2}} + \ldots + \underline{a_n x_{in}}, i = 1, 2, \ldots, n. \tag{5.24}$$

Hence $\hat{y_L} < \acute{y}$. It is to be noted that values of $\hat{y_U}$ in these two cases are the same. Therefore it is clear that $[\acute{y_L}, \hat{y_U}] \subseteq [\hat{y_L}, \hat{y_U}]$. Similarly we can prove $[\acute{y_L}, \hat{y_U}] \subseteq [\hat{y_L}, \hat{y_U}]$ for more than one $\underline{a_j}$. One may note that the values of $\forall \underline{a_j}, j = 0, 1, \ldots, m$ cannot be negative since the boundary values of $\tilde{Y_i}$ and $\tilde{x_{ij}}$ are non-negative.

Now we propose our new method to find the boundaries of the interval regression coefficient of Eq. (5.18) as below. Using the least squares method defined in Section 5.2.4 we find estimates for $\underline{c_j}$'s and $\overline{c_j}$'s of

$$\underline{y_i} = \underline{c_0} + \underline{c_1 x_{i1}} + \ldots + \underline{c_m x_{im}}, \tag{5.25}$$

$$\overline{y_i} = \overline{c_0} + \overline{c_1 x_{i1}} + \ldots + \overline{c_m x_{im}}, \tag{5.26}$$

where $\underline{c_j} = \underline{a_j} - \varepsilon_j, \overline{c_j} = \overline{a_j} - \varepsilon_j$ for $j = 1, 2, ..., m$ which minimizes the following

$$\sum_{i=1}^{N} E_L^2 = \sum_{i=1}^{N} \left(\underline{y}_i - \left(\sum_{0 \le j \le m} \underline{c_j x_{ij}} \right) \right) \tag{5.27}$$

$$\sum_{i=1}^{N} E_R^2 = \sum_{i=1}^{N} \left(\overline{y}_i - \left(\sum_{0 \le j \le m} \overline{c_j x_{ij}} \right) \right). \tag{5.28}$$

One may note that the proposed method can be applied to find the interval estimations for regression coefficients if the boundaries of interval input-output data and center values of interval regression coefficients of Eq.(5.18) are non-negative values. The quality of the approximation can be determined by the average accuracy ratio as mentioned in Section 5.4.1.

This chapter basically discussed developing factor-based models by handling quantitative and qualitative parameters concurrently using the interval function approximation. The interval multivariate regression technique was used to approximate the interval function. Three approaches have been presented to approximate the interval regression coefficients of the interval function.

6 Modeling with the Fuzzy Linguistic Approach

6.1 AN OVERVIEW

The fuzzy linguistic approach is an approximation technique which has been widely applied in many areas (Yager, Goldstein, and Mendels, 1994; Lee, 1996; Chang and Chen, 1994). The technique plays a vital role in the situations where performance values of parameters can only be represented as qualitative terms. These qualitative terms can be interpreted as linguistic values (Mendel, 2006). Here linguistic values are represented in terms of linguistic variables viz. words or sentences. The linguistic value is defined as its label and the semantic (meaning) is represented by the fuzzy set. For a specific application on fuzzy set, the aggregation of linguistic values is an important task if the evaluation procedure deals with the fuzzy linguistic approach.

As mentioned in the literature, there are two major approaches which facilitate undertaking the task of aggregating the linguistic values (Herrera and Herrera-Viedma, 2000). The approaches are the symbolic approach and approximation approach.

In the symbolic approach, linguistic labels are directly computed by considering the nature of the linguistic assessment without considering the semantic (meaning) of the labels. Since linguistic values are approximations it is not needed to carry out the computation with the semantics of the linguistic labels.

The approximation approach deals with the arithmetic of associated membership functions or fuzzy numbers of the fuzzy sets accompanied with the linguistic values. The extended arithmetic operators based on the extension principle are available for the above-mentioned computational procedures. Even though this technique is not as popular as the previous approach, it plays a vital role when the computation cannot be carried out with direct computation on linguistic labels.

In this chapter, we produce factor-based models which have been developed via the above-mentioned two approaches. Several models have been developed concerning two different methods in the symbolic approach. The first method is based on fuzzy linguistic aggregation operators; in that case three operators are chosen as Linguistic Ordered Weighted Average operator (*LOWA*), Linguistic Weighted Average (*LWA*) operator and Majority guided Linguistic Induced Ordered Weighted Average (*MLIOWA*) operator. The second method is based on the fuzzy 2-tuple approach.

The models have been developed based on the approximation approach concerning the situations where merging of the linguistics terms sets of performance values and the importance weights of the parameters is needed.

In the next section, we discuss the model development based on the symbolic approach.

6.2 DEVELOPMENT OF FACTOR-BASED MODELS BASED ON THE SYMBOLIC APPROACH

Before discussing the development procedures of the models, it is important to define the terms linguistic variables and linguistic terms sets which are the basic components in the symbolic approach. First we reproduce the definition of linguistic variable as defined by Zadeh (Zadeh, 1975).

Definition 6.1. (Zadeh, 1975)

A linguistic variable is represented by a quin tuple $(L, H(L), U, G, M)$ where L is the name of the variable; $H(L)$ (or simply H) is the term set of L, i.e., the set of names of linguistic values of L, with each value being a fuzzy variable denoted generically by X and ranging across a universe of discourse U which is associated with the base variable u; G is a syntactic rule for generating the names of the values of L; and M is a semantic rule for associating its meaning with each L, $M(X)$, which is a fuzzy subset of U.

Now let us briefly discuss the linguistic terms set and its properties.

6.2.1 THE LINGUISTIC TERMS SET

In this chapter, we only consider the ordered symmetrically distributed linguistic terms sets. First, the different counting of uncertainty of a particular qualitative parameter/linguistic variable has to be decided. This procedure we called setting the cardinality of the linguistic terms set, i.e. deciding the number of labels in the terms set. For instance, let us assume the qualitative parameter which represents the risk of fire hazards. Here the risk can be broken up into three different risk levels such as Low risk, Moderate risk, and High risk. As to this example the cardinality of the linguistics terms set which represents the risk of fire hazards is three.

The semantic representations of these linguistic labels are given by the fuzzy sets in the interval $[0,1]$. Usually the membership functions accompanied with these fuzzy sets are in the form of triangular membership functions since these functions have been found to be good enough to capture the vagueness of the linguistic assessments (Herrera and Herrera-Viedma, 2000). These fuzzy sets are a triangular fuzzy number which can be represented by a 3-tuple (a,c,b). The first and third points indicate the left and right end point of the domain of the membership function respectively and c indicates the point which attains the maximum membership value, i.e. 1.

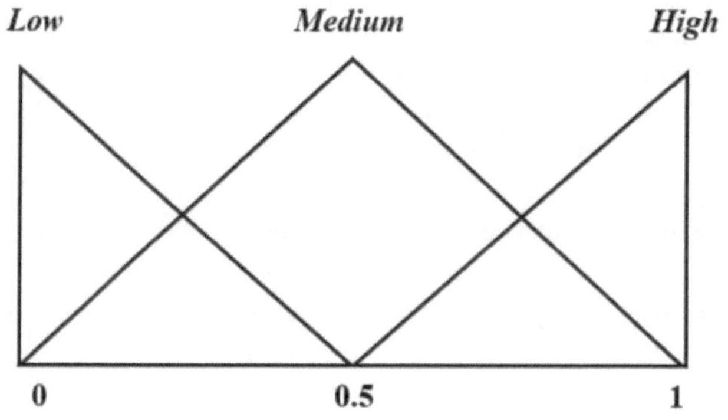

Figure 6.1 Ordered and symmetrically distributed linguistic terms set with cardinality of three.

For example, let us define the semantics of three labels in the linguistic terms set which has been set up to define the risk of fire hazards as follows (see Figure 6.1.):

$s_0 = Low = (0, 0, 0.5)$
$s_1 = Medium = (0, 0.5, 1)$
$s_2 = High = (0.5, 1, 1)$

If the linguistic terms set is chosen as an ordered structure with symmetrical distribution around the modal point 0.5, this linguistic terms set satisfies the properties given below (Herrera and Herrera-Viedma, 2000):

- Negation: $Neg(s_i) = s_j$, $j = T - i$ $(T + 1$ is the cardinality$)$
- Maximization: $\text{Max}(s_i, s_j) = s_i$ if $s_i \geq s_j$.
- Minimization: $\text{Min}(s_i, s_j) = s_i$ if $s_i \leq s_j$.

In the next section, we discuss the development of models based on fuzzy linguistic aggregation operators.

6.3 SYMBOLIC APPROACH - METHOD I - FUZZY LINGUISTIC AGGREGATION OPERATORS

In this method we develop factor-based models by imposing several linguistic aggregation operators such as Linguistic Ordered Weighted Average (*LOWA*), Linguistic Weighted Average (*LWA*), and Majority guided Linguistic Induced Ordered Weighted Average (*MLIOWA*). These models may be categorized into two such as models with non-weighted parameters, i.e. parameters with equal importance and models with weighted parameters, i.e. parameters with unequal importance. In order to develop models with non-weighted parameters, the *LOWA* operator has been used since it combines the non-weighted linguistic information, and for the models with weighted parameters, *LWA* and *MLIOWA* have been considered.

Next we discuss the models based on non-weighted parameters/factors.

6.3.1 MODEL WITH NON-WEIGHTED LINGUISTIC PARAMETERS

Here we present the mechanism to develop models which are based on non-weighted linguistic parameters/factors. In this task we assume that the importance of the parameters involved in the relevant problem is equal. And the aggregation of these parameters is done by the *LOWA* operator. Below we discuss the important facts relating to the *LOWA* operator (Herrera and Herrera-Viedma, 2000).

6.3.1.1 Linguistic Ordered Weighted Averaging (*LOWA*) Operator

This operator possesses properties such as being increasingly monotonous, commutative, and "or-and" and has multiple applications (Herrera, Herrera-Viedma, and Verdegay, 1995; Herrera, Herrera-Viedma, and Verdegay, 1996). Below we reproduce Definition 6.2 as given in (Herrera and Herrera-Viedma, 2000).

Definition 6.2. (Herrera and Herrera-Viedma, 2000)

Let $A = \{a_1, ..., a_m\}$ be a set of labels to be aggregated; then the LOWA operator, ϕ, is defined as

$\phi(a_1, ..., a_m) = W \cdot B^T = \zeta_m \{w_k, b_k, k = 1, ..., m\}$

$= w_1 \odot b_1 \oplus (1 - w_1)$

$\odot \zeta^{m-1} \{\beta_h b_h, h = 2, ..., m\}$,

where $W = [w_1, ...w_m]$, is a weighting vector, such that

(i) $w_i \in [0, 1]$ and, (ii)$\sum_i w_i = 1, \beta_h = \dfrac{w_h}{\sum_2^m w_k}, h = 2, ..., m$ and

(iii)$\{b_1, ..., b_m\}$

is a vector associated to A, such that,

$B = \sigma(A) = \left\{a_{\sigma(1)}, ..., a_{\sigma(n)}\right\}$ in which, $a_{\sigma(j)} \leq a_{\sigma(i)}\ \forall i \leq j$, with σ being a permutation over the set of labels A. ζ^m is the convex combination operator of m labels and if $m = 2$, then it is defined as

$\zeta^2 \{w_i, b_i, i = 1, 2\} = w_1 \odot s_j \oplus (1 - w_1) \odot s_i = s_k,\ s_j, s_i \in S\ (j \geq i)$

such that $k = \min\{T, i + round\,(w_1 \cdot (j - i))\}$, where "round" is the usual round operation, and $b_1 = s_j, b_2 = s_i$.

If $w_j = 1$ and $w_i = 0$ with $i \neq j \forall i$, then the convex combination is defined as $\zeta^m \{w_i, b_i, i = 1, ..., m\} = b_j$.

6.3.1.2 The Concept of Fuzzy Majority in the Aggregation of LOWA Operator

In order to aggregate the labels, one may need to find the weight vector which reflects the fuzzy majority. As to the definition given above this vector is denoted as w and these can be evaluated using fuzzy linguistic quantifiers (Gabriella and Yager, 2006). The meaning or semantic of a linguistic quantifier can be interpreted through fuzzy sets (Zadeh, 1975). In the following, the method to find the weights by means of a non-decreasing proportional quantifier as defined in (Herrera-Viedma, Pasi, Lopez-Herrera, and Porcel, 2006) is reproduced:

The weights by means of a non-decreasing proportional quantifier QF is given by:
$w_i = Q(i/n) - Q((i-1)/n),\ i = 1, ..., n,$
where the membership function of QF is:

$$QF(i/n) = \begin{cases} 0 & \text{if } r < p, \\ \dfrac{(r-p)}{(q-p)} & \text{if } p \le r \le q, \\ 1 & \text{if } r > q, \end{cases} \tag{6.1}$$

with $p, q, r \in [0, 1]$. Most frequently used linguistic quantifiers are "Mean", "As many as possible", "At least half", "Most", etc. The Mean quantifier QF_{Mean} is a pure averaging quantifier where the weights are evaluated by the equation
$w_i = \frac{1}{n}$, for all $i = 1, ..., n$
The membership function for the QF_{Mean} is (Herrera and Herrera-Viedma, 1997):

$$QF_{Mean}(N) = \frac{N}{n}, N = 1, ..., n. \tag{6.2}$$

The parameter (p, q) of the quantifiers "Most", "At least half", "As many as possible" are $(0.3, 0.8)$, $(0, 0.5)$, and $(0, 1)$, respectively (see Figure 6.2).

For example, assume that a set of three labels is given to aggregate using the LOWA operator. In this task, the first step is to find the weight vector w_i with respect to a relevant linguistic quantifier. One may choose the "Most" linguistic quantifier. As mentioned above the parameter for the Most quantifier is $(0.3, 0.8)$.

Here three weights have to be found, since there are three labels, i.e. $i = 3$, when $i = 1$,
$w_1 = QF(1/3) - QF((1-1)/3),$
$w_1 = QF(1/3) - QF(0/3) = QF(0.33) - QF(0).$

Likewise, for $i = 2$ and $i = 3$ we get

$w_2 = QF(2/3) - QF(1/3) = QF(0.67) - QF(0.33)$

$w_3 = QF(3/3) - QF(2/3) = QF(1) - QF(2/3) = QF(1) - QF(0.67).$

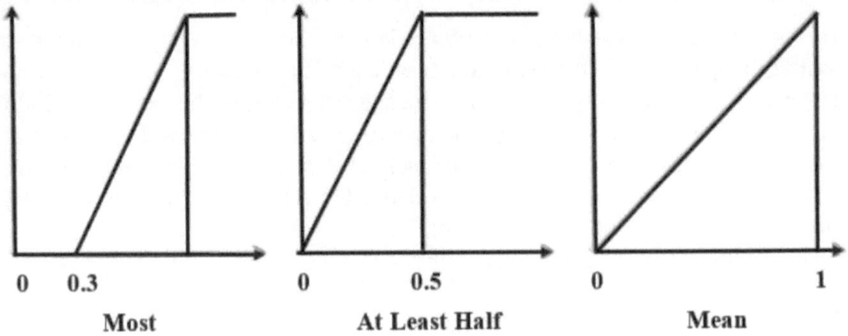

Figure 6.2 Fuzzy Linguistic Quantifiers. (Reprinted from *Novel Fuzzy Linguistic Based Mathematical Model to Assess Risk of Invasive Alien Plant Species,* 59, Peiris, H.O.W., Chakraverty, S., Perera, S.S.N., and Ranwala, S.M.W. 326-339, Copyright (2017), with permission from Elsevier.)

The values for $QF(0.33), QF(0.67), QF(1)$, and $QF(0)$ can be evaluated from Eq. (6.1).

Now take $QF(0.33)$. Here the value 0.33 is in between 0.3 and 0.8, so the membership value is $\frac{0.33-0.3}{0.8-0.3}$=0.06.
Following the same procedure, values for the $QF(0.67), QF(1), QF(0)$ are evaluated as 0.74, 1, 0 respectively. Therefore the values of the weights are:

w_1=$QF(0.33)-QF(0)$=0.06-0=0.06

w_2=$QF(0.67)-QF(0.33)$=0.74-0.06=0.68

w_3=$QF(1)-QF(0.67)$=1-0.74=0.26

Let us now discuss an example of how to apply the LOWA operator to aggregate the linguistic parameters in the specific problem given below:

Example 6.1 A company wants to hire a suitable candidate by considering the areas such as level of knowledge, experience, and qualifications. Assume two candidates have applied for this position. A panel of senior managers rates each candidate with respect to each area mentioned above using a uniformly and symmetrically distributed linguistic terms set consisting of three labels such as $s_0 = Low, s_1 = Medium$, and $s_3 = High$ (similar to Figure 6.1). The indexes of the labels Low, Medium, and High are 0,1,2 respectively and the cardinaility of the linguistic terms set is 3.

Table 6.1
Performance values of candidates

Candidate No	Knowledge	Experience	Qualification
1	Medium	Medium	High
2	High	Low	Medium

Table 6.1 shows how each candidate scored with respect to each area.

Now let us assume that the panel may evaluate each candidate by aggregating the performance value in each area by the LOWA operator with "Most" quantifier.
In this computation, as we mentioned before, the first step is to find the weight vector by means of the chosen quantifier. Since the aggregation undergoes three linguistic labels with Most quantifier, the values of the three weights are similar to the values that have been evaluated in the previous section.
Therefore the weights are w_1=0.06, w_2=0.68, w_3=0.26

Let us first consider Candidate 2 with performance values (High, Low, Medium) with respect to each area as given in Table 6.1.
The final score for Candidate 2 may be evaluated using the LOWA operator with the relevant weights as follows:

Step 1: Prepare the performance values in descending order.
Then we have (High, Medium, Low)

Step 2: First consider the pair (Medium, Low). Then evaluate
$k_2=\min\left\{T-1, 1+r\left(\frac{w_1}{w_2+w_1}\times(j-i)\right)\right\}$
where j denotes the index of larger performance value and i denotes the index of smaller performance value. The T denotes the cardinality of the given linguistic terms set. According to the given facts $T=3$, $j=1$, and $i=0$. Then we get
$k_2=\min\left\{2, 1+r\left(\frac{0.06}{0.94}\times 1\right)\right\}$=1=Medium.

Step 3: For the pair (High, Medium) which consists of a performance value that remains in the triplet and a resultant value in k_2 evaluate

$$k_3=\min\left\{T-1, 1+r\left(\frac{w_2}{w_2+w_3}\times(j-i)\right)\right\}.$$

Here $T=3$, $j=2$, and $i=1$. Then we get $k_3=\min\left\{2, 1+r\left(\frac{0.68}{0.94}\times(j-i)\right)\right\}$=2 =High.

Then the aggregated performance value of Candidate 2 is *High*.

Next we discuss the development procedure of factor-based models using LOWA.

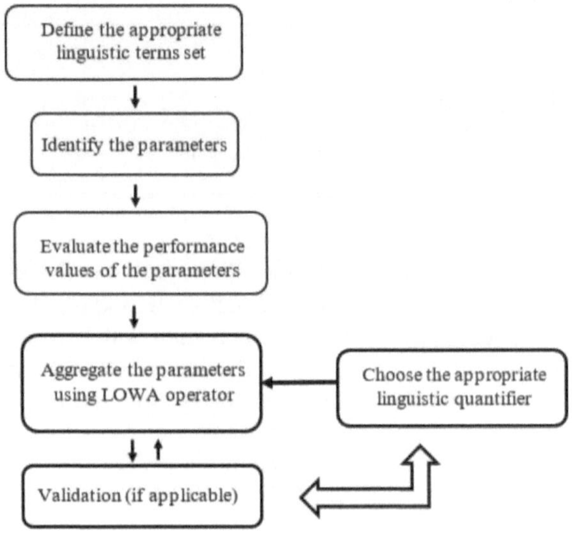

Figure 6.3 Structure of the model – Non-weighted parameters.

6.3.1.3 Setting Up Model I-Non-weighted

Here we focus on developing a factor-based model based on the LOWA operator. This model may be applied in specific problems which satisfy the following conditions:

- The performance value of each and every parameter should be in the qualitative/linguistic form where the labels are defined in a uniformly and symmetrically distributed linguistic terms set.
- The parameters are equally important for the main objective of the problem.

Figure 6.3 depicts the structure of the model constructed based on the conditions mentioned above. Development steps of the model are briefly explained next.

Step 1: First, the qualitative/linguistic parameters which describe the specific problem should be identified.

Step 2: After investigating the behavior of the selected parameters, the cardinality of the linguistic terms set should be defined.

Step 3: Evaluate the performance values of the parameters based on the linguistic terms set which is defined in Step 2.

Step 4: By choosing an appropriate linguistic quantifier ("Mean", "Most", "At least half", "As many as possible", etc.) aggregate performance values using LOWA operator.

Step 5: Perform validation for the output values if applicable.

Step 6: If the output is need to be revised then redirect to Step 5 and perhaps change the linguistic quantifier appropriately.

In the next subsection, we discuss the factor-based models for weighted parameters.

6.3.2 MODEL WITH WEIGHTED PARAMETERS

There are situations in which the importance of a parameter for the problem is different from parameter to parameter. As such, importance weights of the parameters need to be considered in the aggregation process.
Therefore, the parameters with importance weights have to undergo two aggregation phases as (Herrera and Herrera-Viedma, 2000):

- The aggregation of importance weights of model parameters.
  In this phase, the importance weight of a parameter is aggregated with its corresponding performance value.
- The aggregation of weighted information.
  At this phase, all the performance values of parameters after aggregating with the importance weights are again aggregated to evaluate the overall performance.

In this task, we consider the Linguistic Weighted Averaging operator and Majority guided Induced Ordered Weighted Average operator which are specifically designed for aggregating weighted linguistic information (Herrera and Herrera-Viedma, 1997; Herrera-Viedma, Pasi, et al., 2017; Lopez-Herrera and Porcel, 2006; Tao, et al., 2015; Yager, 1980b; Zhang, et al., 2016; Yager 2001). First let us reproduce the Linguistic Weighted Averaging operator as stated in (Herrera and Herrera-Viedma, 1997):

Linguistic Weighted Averaging (LWA) Operator

Definition 6.3. The aggregation of the set of weighted individual opinions, $\{(w_1,a_1),...,(w_n,a_n)\}$, according to the Linguistic Weighted Averaging (LWA) operator is defined as
$(w_E,a_E) = LWA\left[(w_1,a_1),...,(w_n,a_n)\right]$,
where the importance weights of the group opinion, w_E, is obtained as
$w_E = \phi_Q(w_1,...,w_n)$.
And the opinion of the group a_E, is obtained as

Table 6.2
Importance Weights of parameters of candidate evaluation

Parameter	Importance Weight
Knowledge	High
Experience	High
Qualifications	Medium

$a_E = f\left[g\left(w_1,a_1\right),g\left(w_n,a_n\right)\right]$,
where $f = \phi_Q$ and $g \in \left\{LC_1^{\rightarrow}, LC_2^{\rightarrow}, LC_3^{\rightarrow}\right\}$.

Here $LC^{\rightarrow}$ is a linguistic conjunction function which is monotonically non-decreasing in the weights and satisfies the properties required for any transformation function g. The available conjunction functions are classical MIN and classical MAX operators. The classical MIN operator is defined as in (Herrera and Herrera-Viedma, 1997):
$LC_1^{\rightarrow}\left(w,a\right) = \text{MIN}\left(w,a\right)$

Example 6.2 Here, other than the facts presented in Example 6.1, the importance weights of the three parameters are also assigned by the panel. Table 6.2 presents weights assigned for the three parameters: knowledge, experience, and qualifications. One may note that the importance weights have also been assigned based on the same linguistic terms set used to evaluate the performance values of the parameters.

As we mentioned above, the first task is to combine the performance value of a parameter with its corresponding importance weight. Let us extract the performance values of Candidate 1 with respect to each parameter from Table 6.1 and paired with the corresponding weights as follows:

$$\left\{\left(High, Medium\right), \left(High, Medium\right), \left(Medium, High\right)\right\}.$$

The first element in a particular pair in the above denotes the importance weight and the second element denotes the performance value of the object. To combine the two elements in a pair we use a classical min operator as follows:

$$LC_1^{\rightarrow}\left\{\left(High, Medium\right), \left(High, Medium\right), \left(Medium, High\right)\right\}$$
$$= \left\{min\left(High, Medium\right), min\left(High, Medium\right), min\left(Medium, High\right)\right\}$$
$$= \left\{\left(Medium\right), \left(Medium\right), \left(Medium\right)\right\}.$$

Now one may see that each pair has been reduced to one label. In order to evaluate the overall performance, one may need to follow up with the same process as we discussed in Example 6.1.
We discuss the developing steps of the model based on the LWA operator next.

Setting up Model II-Weighted - *LWA*

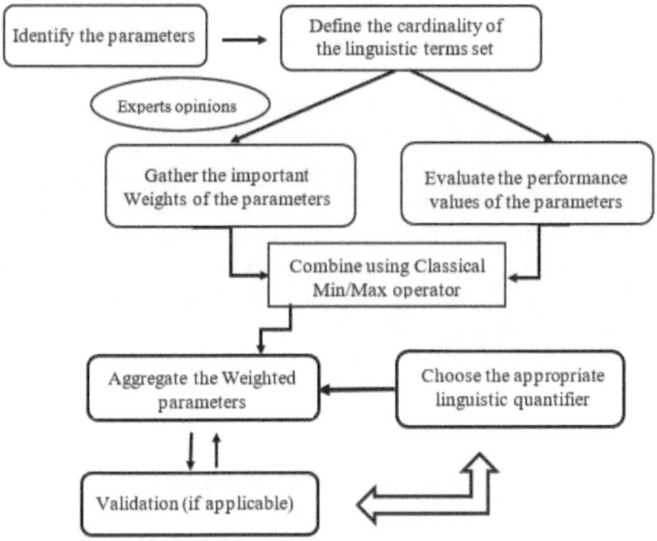

Figure 6.4 Structure of Model II – Weighted parameters – LWA operator.

Here we discuss the methodology to develop a factor-based model using the LWA operator. The following conditions have to be satisfied in order to use this model for a problem:

- The performance value of each and every parameter should be in the form qualitative/linguistic where the labels are defined in a uniformly and symmetrically distributed linguistic terms set.
- The parameters should have different importance weights where those weights are collected from the same linguistic terms set defined for the performance values for the parameters mentioned in the first condition.

Figure 6.4 illustrates the developing steps of Model II.
Below we briefly discuss steps of Model II as given in Figure 6.4.

Step 1 and Step 2: Observe that the first two steps are same as Model I.

Step 3: In this step the performance values of the parameters are gathered using the linguistic terms set defined in Step 2 and also the importance weights based on the experts opinions using the same linguistic term set are obtained.

Step 4: Here the importance weight of each parameter is combined with the corresponding performance value of the parameters using the classical Min/Max operator.

Step 5: The weighted parameters obtained from the previous step are aggregated by choosing an appropriate linguistic quantifier.

Step 6: In this step, the output is validated. If there are any considerable deviations then the process is redirected to the previous step.

One may note that the Majority guided Linguistic Induced Ordered Weighted Average (*MLIOWA*) operator has been specially designed to overcome the limitations that occur in the majority guided OWA operators such as *LOWA*, *LWA*, etc. (Dong and Herrera-Viedma, 2015; Herrera-Viedma, et al., 2017; Lopez-Herrera and Porcel, 2006; Tao, et al., 2015; Zhang, et al., 2016).
Next we reproduce the *MLIOWA* operator as defined in (Herrera-Viedma, et al., 2006).

Definition 6.4. (Herrera-Viedma, et al., 2006) A weighted *MLIOWA* operator of dimension n is a function
$\Phi_Q^I : (S \times S)^n \longrightarrow S$, defined according to the expression

$$\Phi_Q^I((I_1, p_1), \ldots, (I_n, p_n)) = \Phi_Q((u_1, p_1), \ldots, (u_n, p_n))$$

in which
1. The order-inducing values are obtained from the linguistic importance degrees associated with the values to be aggregated as

$$u_i = \frac{\text{sup}_i + ind\,(I_i)}{2}$$

$$\text{sup}_i = \Sigma_{j=1}^{n} \text{sup}_{ij} | \text{sup}_{ij} = \begin{cases} 1 \ \& \text{ if } |ind\,(I_i) - ind\,(I_j)| < \sigma \\ 0 \ \& \text{ otherwise,} \end{cases}$$

where I_i is the linguistic importance degree of the value p_i to be aggregated and $ind()$ generates the index of the label.

2. The weighting vector is evaluated as

$$w_i = Q\left(\frac{u_{\sigma(i)}}{n}\right) / \Sigma_{j=1}^{n} Q\left(\frac{u_{\sigma(j)}}{n}\right)$$

Let us recall Example 6.2 and evaluate the overall performance of the first candidate with respect to the three parameters: Knowledge, Experience, Qualifications. Below we briefly explain the steps relating to this evaluation:

Calculating sup_i This calculation is based on the importance weights. As to this example, $i = 3$ since there are three importance weights. For each importance weight the sup_i needs to be calculated. For example, take the importance weight of the parameter "Knowledge", i.e. "High". First calculate the $sup_{1,1}$ for the parameter "Knowledge" w.r.t itself. Since both labels are the same then we get $|2 - 2| < 1$ and

Table 6.3
Values of sup_i and u_i

Parameter	Importance weight	sup_i	u_i
Knowledge	*Very High*	2	2
Experience	*Very High*	2	2
Qualifications	*Medium*	1	1

$sup_{1,1} = 1$ (Here α is taken as 1). For $sup_{1,2}$ w.r.t. "Experience" we get the same value as $sup_{1,1}$, i.e. 1.

Then calculate $sup_{1,3}$ w.r.t parameter "Qualification". Here the importance weight of the parameter "Qualification" is "Medium" and the index is 1. Hence we get $|2-1| < 1$. Then $sup_{1,3} = 0$. Therefore the $sup1 = sup_{1,1} + sup_{1,2} + sup_{1,3} = 1+1+0=2$ is obtained for the parameter "Knowledge". Following the same procedure, the $sup2$ and $sup3$ can be evaluated.

The order-inducing value u_1 for the parameter "Knowledge" is calculated as follows:

Here the $sup1=2$ and $ind(I_1)$ is 2 since the label of the importance weight is "High."

Therefore $u_1 = \dfrac{2+2}{2} = 2$. Following the same procedure order-inducing values for the remaining parameters can be found. The values of $supi$ and ui obtained for each parameter are presented in Table 6.3.

Calculating w_i Let us now evaluate the weight vector w_i by considering "At least half" as the quantifier which is defined as $(0, 0.5)$.
First arrange the importance weights in Table 6.3 in ascending order. Then the weights w_1, w_2 and w_3 correspond to the importance weights "Medium", "High", "High" respectively.
Here $w_1 = Q\left(\frac{u_{\sigma(i)}}{n}\right) / \Sigma_{j=1}^{n} Q\left(\frac{u_{\sigma(j)}}{n}\right) = Q\left(\frac{1}{3}\right) / \left(Q\left(\frac{1}{3}\right) + Q\left(\frac{2}{3}\right) + Q\left(\frac{2}{3}\right)\right)$
$= \dfrac{0.333}{1.333}$. The values for $Q\left(\frac{1}{3}\right), Q\left(\frac{2}{3}\right)$ can be obtained by applying Eq. (6.1) by taking $p = 0$ and $q = 0.5$.

Thus, the value for w_2 and w_3 is $\dfrac{0.5}{1.333}$.

Aggregation The final task is to aggregate the weighted parameters.
Now assume that we need to aggregate the following pairs with respect to the weights obtained.

$\Phi_{MLIOWA}\{(High, Medium), (High, Medium), (Medium, High)\}$.

First rearrange the pairs in ascending order considering the importance weights, i.e. the first element of the pair. Then we get

$$\Phi_{MLIOWA}\{(Medium, High), (High, Medium), (High, Medium),\}.$$
$$=\text{round}\left(2 * \frac{0.333}{1.333} + 1 * \frac{0.5}{1.333} + 1 * \frac{0.5}{1.333}\right)$$
$$=\text{round}(1.25)=1=\text{Medium}.$$

In the following, we discuss the developing steps of the model based on the MLIOWA operator.

Setting up Model III-Weighted - *MLIOWA*

The conditions that have to be satisfied in order to use this model are the same as the conditions set up for the Model II-*LWA*.

Figure 6.5 illustrates the developing steps of the Model III−*MLIOWA*. Below we briefly discuss the steps involved in this model:

Steps 1 to 3: The first three steps are similar to Model I−*LOWA*.

Step 4: The sup_i value for the each importance weight obtained in Step 3 is calculated.

Step 5: The weight vector w_i is evaluated with respect to the importance weight of each parameter. For this task, the appropriate linguistic quantifier has to be chosen.

Step 6: In this step, combining of importance weights and performance values of parameters and overall aggregation are carried out consequently.

In the following section, we discuss development of factor-based models based on the fuzzy 2-tuple representation.

6.4 SYMBOLIC APPROACH - METHOD II - FUZZY 2-TUPLE

In the previous section, we have discussed three models based on the operators *LOWA*, *LWA*, and *MLIOWA*. One may note that these operators are executed based on the extension principle and symbolic methods. Also, some steps are followed with rounding off the operator which leads to loss of information where the results are expressed in a discrete domain. Moreover, this issue can affect the preciseness of the final result.

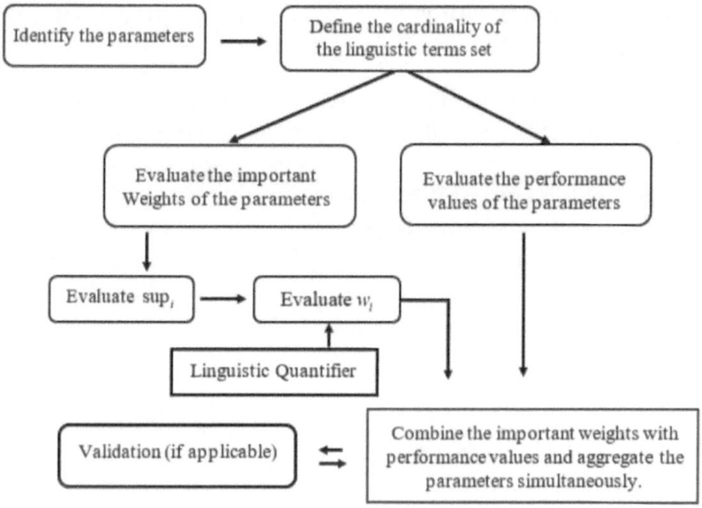

Figure 6.5 Structure of Model III – Weighted parameters – MLIOWA operator.

The fuzzy 2-tuple representation has been developed to overcome the limitations mentioned above (Herrera and Martinez, 2000; Dong, et al., 2015; Dong and Herrera-Viedma, 2015; Lee and Chen, 2015; Wei, et al., 2018; Li, et al., 2017; Li, Dong, Herrera, et al., 2017). Here the linguistic information is stored by means of 2-tuples which are composed of a linguistic term and a numeric value assessed in $[0.5, 0.5)$. The numeric value facilitates continuous representation of the linguistic information on the given domain without loss of information.

In this task, we develop two factor-based models based on 2-tuple representation. In the first model, the parameters are considered as equally important where in the second model those parameters are considered as unequally important.

In the following section, we present some theoretical aspects of 2-tuple linguistic approach.

6.5 FUZZY 2-TUPLE LINGUISTIC REPRESENTATION

The 2-tuple representation is a continuous linguistic representation which does not exactly match any linguistic term.
It represents the linguistic information by means of a pair of values (s, α), where s is a linguistic term and α is a numeric value representing the symbolic translation.

6.5.1 SYMBOLIC TRANSLATION OF FUZZY 2-TUPLE LINGUISTIC REPRESENTATION

The symbolic translation of the fuzzy 2-tuple linguistic representation function has been reproduced in Def. 6.5 as stated in (Herrera and Martinez, 2000).

Definition 6.5. Let $S = \{s_0, ..., s_g\}$ be a linguistic term set and $\beta \varepsilon [0, 1]$ a value representing the result of a symbolic aggregation operation; then the 2-tuple that expresses the equivalent information to β is obtained with the following function:

$$\Delta : [0, g] \rightarrow S \times [-0.5, 0.5)$$

$$\Delta(\beta) = (s_i, \alpha) \text{with} \begin{cases} s_i & i = round(\beta) \\ \alpha = \beta - i & \alpha \varepsilon [-0.5, 0.5) \end{cases} \tag{6.3}$$

where round$(\cdot)$ is the usual round operation, s_i has the closet index label to "β," and "α" is the value of the symbolic translation.

6.5.2 FUZZY 2-TUPLE AGGREGATION OPERATORS

In this representation, the linguistic information is in the form of an ordered pair as mentioned in the previous subsection. As mentioned in the literature, the arithmetic mean and weighted average are the basic 2-tuple aggregation operators developed based on classical aggregation operators (Herrera and Martinez, 2000; Dong and Herrera-Viedma, 2015).

Below, we reproduce the arithmetic mean operator equivalent to the linguistic 2-tuple as in (Herrera and Martinez, 2000).

Definition 6.6. (Herrera and Martinez, 2000) Let $x = \{(r_1, \alpha_1), ..., (r_n, \alpha_n)\}$ be a set of 2-tuples, then the 2-tuple arithmetic mean $\bar{x}^e$ is computed as

$$\bar{x}^e = \Delta\left(\sum_{i=1}^{n} \frac{1}{n} \Delta^{-1}(r_i, \alpha_i)\right) = \Delta\left(\frac{1}{n} \sum_{i=1}^{n} \beta_i\right) \tag{6.4}$$

This operator facilitates aggregating the mean of a set of linguistic 2-tuples without any loss of information. In the weighted average operator, the grade of importance weight has been incorporated with each value x_i. The weighted average operator equivalent to the linguistic 2-tuple is reproduced in Def 6.7 as stated in (Herrera and Martinez, 2000).

Definition 6.7. (Herrera and Martinez, 2000) Let $x = \{(r_1, \alpha_1), ..., (r_n, \alpha_n)\}$ be a set of 2-tuples,and $W = \{w_1, ..., w_n\}$ be their associated weights. Then the 2-tuple

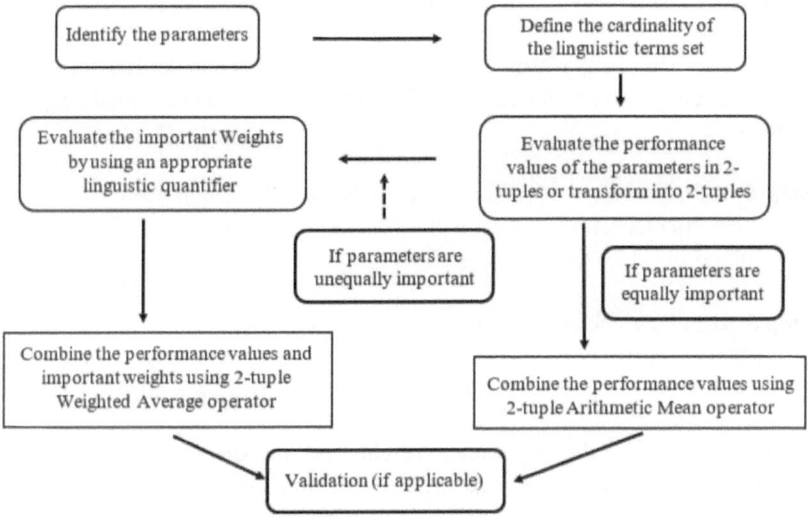

Figure 6.6 Structure of Model IV.

weighted average $\bar{x}^e$ is

$$\bar{x}^e = \Delta\left(\frac{\sum_{i=1}^{n}(r_i, \alpha_i)\cdot w_i}{\sum_{i=1}^{n} w_i}\right) = \Delta\left(\frac{\sum_{i=1}^{n}\beta_i \cdot w_i}{\sum_{i=1}^{n} w_i}\right). \tag{6.5}$$

6.5.3 SETTING UP MODEL IV

Here we present the methodology to develop a factor-based model where the performance values of the parameters are in the form of linguistic 2-tuples. In order to use this model, the following conditions should be satisfied:

- The performance value of each and every parameter should transform into the linguistic 2-tuples where the labels are defined in a uniformly and symmetrically distributed linguistic terms set.
- If parameters have different importance degrees, those weights should be assigned using a linguistic quantifier.

Based on the above conditions, the model has been developed as presented in Figure 6.6. We briefly explain the steps of the model as follows:

Steps 1 to 2: The first two steps are similar to Model III.

Step 3: Evaluate the performance values of the parameters and transform into 2-tuples. Depending on the problem, the performance values may be directly expressed as 2-tuples.

Step 4: If the parameters are considered to be equally important for the problem then the performance values are aggregated using the 2-tuple arithmetic mean operator as defined in Eq. (6.4).

Step 5: If the parameters are considered to be unequally important for the problem as a first step, the importance weights of the parameters have to be evaluated using an appropriate linguistic quantifier.

Step 6: In this step, the importance weights and the performance values are aggregated using the weighted average operator as defined in the Eq. (6.5).

Step 7: Validation can be carried out to obtain a significant output.

To give a clearer explanation of the aggregation taking a part in the above steps, below we discuss Example 6.3.

Example 6.3 This is the same example mentioned in Example 6.1; however, we consider only the the performance values of the first candidate as given in Table 6.4. One may note that, in the third column, the performance values are converted into the 2-tuples. To transform the performance values in the second column which are in the single labels into 2-tuples, one may need to add the symbolic translation as zero.

Let us now aggregate the performance values of the three parameters using the 2-tuple arithmetic mean operator (see Eq.(6.4)) to evaluate the candidates' performance by assuming parameters are equally important.

As the first step the β_i values for the parameters need to be calculated. The value of β_i is the summation of the index of the label s_i and the symbolic translation α. As to the given information, α is zero then β_i values are (1,1,2) where 1,2 represent the index of the labels "Medium", "High" respectively.

Using the 2-tuple arithmetic mean operator, we get

$$\Delta\left(\frac{1}{n}\sum_{i=1}^{n}\beta_i\right) = \left(\frac{1+1+2}{3}\right) = 1.33.$$

The remaining step is to match the 2-tuple which represents the value 1.33. According to Eq. (6.3), the rounded off value of β_i, i.e. 1.33, is 1 and then s_i=Medium and $\alpha = 1.33 - 1$=0.33. Therefore the performance of the first candidate is (Medium,0.33).

Table 6.4
Performance values of first candidate

Parameter	Performance Value	2-tuple
Knowledge	Medium	(Medium,0)
Experience	Medium	(Medium,0)
Qualification	High	(High,0)

Now assume that the three parameters are unequally important to evaluate the candidate's performance. In that case, a linguistic quantifier needs to be chosen to find the importance weights. For instance, take the "Most" quantifier with the parameter (0.3,0.8).
Then the weight vector is $(0.06, 0.68, 0.26)$.

After arranging the β_i values in ascending order, i.e. (1,1,2), apply the 2-tuple weighted average operator as follows:

$$\Delta\left(\frac{\sum_{i=1}^{n}\beta_i w_i}{\sum_{i=1}^{n} w_i}\right) = \frac{(1\times 0.06 + 1\times 0.68 + 2\times 0.26)}{1} = 1.26.$$

The remaining step is to match the 2-tuple which represents the value 1.33. According to Eq. (6.3), the rounded off value of β_i, i.e. 1.33, is 1 and then s_i=Medium and $\alpha = 1.33 - 1 = 0.33$. Therefore the performance of the first candidate is (Medium,0.33).

Lastly, the 2-tuple which matches with the value 1.26 is evaluated using Eq. (6.3). Here the rounded off value of β_i, i.e. 1.26, is 1 and then s_i=Medium and $\alpha = 1.26 - 1 = 0.26$. Therefore the performance of the first candidate is (Medium,0.26).

In the next section, we focus on factor-based models based on the approximation approach.

6.6 APPROXIMATION APPROACH

In the previous two methods, those evaluations are based on direct computation of the indexes of labels of the linguistic terms. However, there are situations where experts do not agree to give their opinions on the importance weights of the parameters based on the terms set that we provide. In that case, they might use a terms set where its cardinality and the semantic(meaning) of the labels are different compared to the terms set used to evaluate the performance values of the parameters.
Due to the differences between cardinalities and semantics, direct computation of labels is un-executable; however, the evaluation is possible with the semantic(meaning)

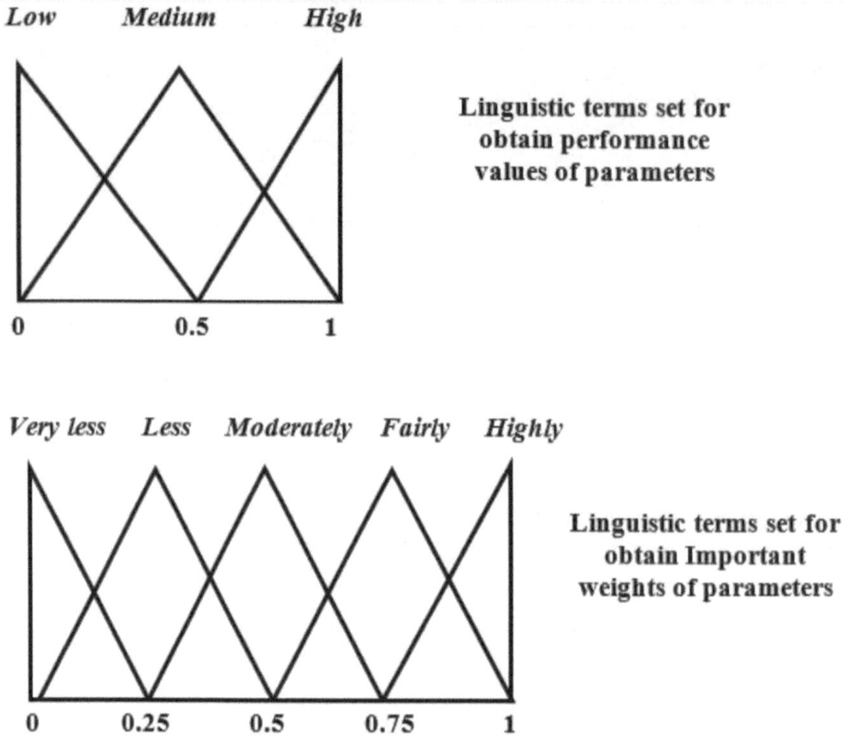

Figure 6.7 Two different linguistic scales defined for performance values and importance weights of parameters.

of the labels. Figure 6.7 depicts two different linguistic terms sets used to evaluate the performance values and the importance weights of the parameters in a given problem.

In such a situation, we need a proper mechanism to merge the linguistic scale to aggregate the weighted parameters. Therefore, following the algorithm proposed by Lee (Lee, 1996) we have developed a factor-based model which can be applied for problems where the merging of two linguistic terms sets is needed. In the following, we discuss the method to merge the terms sets to evaluate the final aggregation of the given parameters.

6.6.1 METHOD TO MERGE LINGUISTIC TERMS SET

In this task, we assume that the judgments of importance weights and performance values of the qualitative/linguistic parameters are gathered from two different linguistic scales. Another fact that has been considered here is that the parameters may be present as main and sub parameters. Following up this merging technique, a numerical value is generated as the final aggregation of the weighted parameters.

The components of this technique are defined as follows:

GIW_{M_j}- Grade of importance weight of jth main parameter. (numerical value)

$(PM_j)_i$- Performance value of jth main parameter of ith alternative.(fuzzy number of corresponding linguistic term)

$\left(PS_{m_j}\right)_i$- Performance value of mth sub parameter of jth main parameter of ith alternative.(fuzzy number of corresponding linguistic term)

$GIW_{S_{m_j}}$- Grade of importance weight of mth sub parameter of jth main parameter. (numerical value)

$SAGG_{m_j}$- Aggregation of $GIW_{S_{m_j}}$ and $\left(PS_{m_j}\right)_i$.(numerical value)

$(MAGG_j)_i$-Performance value of jth main parameter by aggregating its weighted sub parameters, i.e. aggregation of $SAGG_{m_j}$. (numerical value)

$OAGG_i$- Overall aggregation of weighted parameters of ith alternative, i.e. aggregation of $(MAGG_j)_i$. (numerical value)

$DOAGG_i$ Defuzzified value of the $OAGG_i$.

Let us now focus on calculating the components $SAGG_{m_j}$,$(MAGG_j)_i$, $OAGG_i$, and $DOAGG_i$ mentioned above.

1. Calculating $SAGG_{m_j}$
   For a particular i^{th} alternative $SAGG_{m_j}$ is defined as

$$SAGG_{m_j} = d\left(GIW_{S_{m_j}} \bigotimes (\left(PS_{m_j}\right)_i\right) \tag{6.6}$$

   where d denotes the centroid method which is one of defuzzification methods and $\otimes$ denotes the usual multiplication of normal triangular fuzzy numbers.
   For example, if $GIW_{S_{m_j}}$=(0 0.5 1) and $\left(PS_{m_j}\right)_i$=(0.5 0.75 1) then

   d($GIW_{S_{m_j}} \bigotimes (\left(PS_{m_j}\right)_i$)= ((00.51) $\bigotimes$ (0.50.751))=d(00.3751)=$\left(\frac{0+0.375+1}{3}\right)$
   =0.458.
2. Calculating $(MAGG_j)_i$
   Here we evaluate the performance value of the jth main parameter if it has sub parameters; otherwise the value is $(PM_j)_i$.

Assume cardinality of the linguistic scale considered to obtain the performance values of the n parameters and $\mu_n(x)$ is the membership function of the nth label. The intersection of each $SAGG_{m_j}$ with each membership function with respect to each label in the linguistic scale is defined as $L_m(1,2,...,k)$, $k = 1,2,...,n$.
For example, if $SAGG_{m_j}$=0.45 and the linguistic scale is the first scale in Figure 6.7 then $L_m(1,2,...,k) = (0,0.2,0.8,0,0)$.
Then $(MAGG_j)_i$=$\Sigma_m L_m(1,2,...,k)$ where $\Sigma_{k=1}^{n}(MAGG_j)_i$=1.

3. $OAGG_i$
The overall aggregation is defined as

$$OAGG_i = GIW_{M_j} \otimes (MAGG_j)_i = (OAGG_1, OAGG_2, ..., OAGG_n). \qquad (6.7)$$

4. $DOAGG_i$
Here $DOAGG_i$ is obtained by defuzzifying $OAGG_i$ as follows:

$$DOAGG_i = \frac{\sum_{k=1}^{n} GV(k) \otimes OAGG_i}{\sum_{k=1}^{n} OAGG_i} \qquad (6.8)$$

where $GV(k)$ is the values of the linguistic terms set defuzzified by the centroid method.

6.6.2 SETTING UP MODEL V

Here we discuss the method to develop the model based on approximation. In order to apply this model, the following conditions should be satisfied.

- The performance value of each and every parameter should be in the form qualitative/linguistic where the labels are defined in a uniformly and symmetrically distributed linguistic terms set.
- The qualitative parameters may be categorized into main and sub parameters.
- The linguistic terms set which is defined to evaluate the importance weights is different from the linguistic terms set defined in the first condition by the cardinality and semantics of the labels.
- The importance weights can be gathered as a group decision or individual opinions of experts.
- The final aggregation is based on the linguistic term defined for the performance values for the parameters mentioned in the first condition.

Figure 6.8 illustrates the structure of Model V. Below, we briefly discuss the development steps of the model:

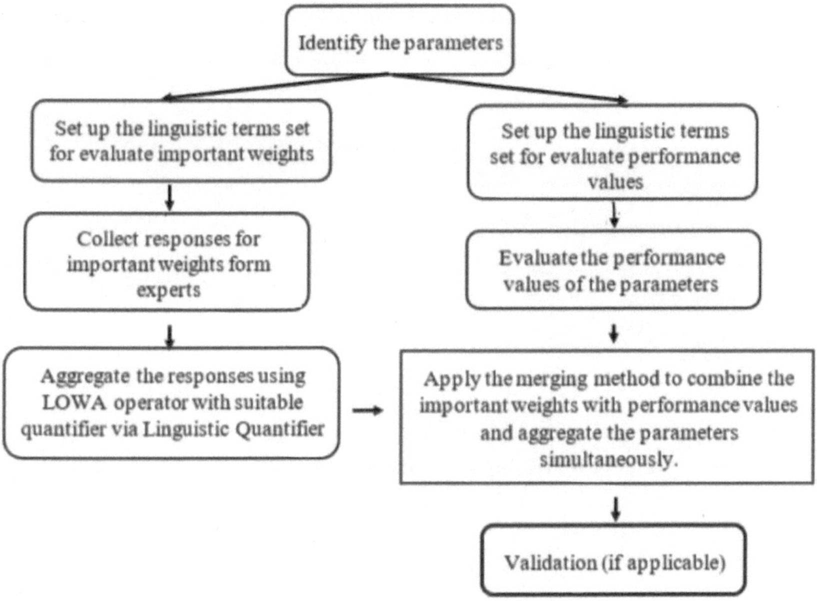

Figure 6.8 Structure of Model V – Merging of linguistic terms sets

Step 1: The qualitative/linguistic parameters which describe the specific problem need to be investigated.

Step 2: Basically, the cardinality and semantics of the labels of the linguistic terms set has to be set up. Based on expert's opinions, the linguistic terms set for importance weights has to be defined.

Step 3: The performance values of the parameters are gathered using the linguistic terms set constructed in Step 2. The experts' opinions on importance weights are collected as a group decision or individual opinions for each parameter. This can be done by the pairwise comparison method which has been discussed in Chapter 4.

Step 4: If the experts' opinions are gathered as individual opinions then the opinions with respect to each parameter are aggregated using the *LOWA* operator with an appropriate linguistic quantifier. Moreover, the individual opinions are given as pairwise comparisons; then the *LOWA* operator has to be applied twice. In that case, after aggregating the opinions, to choose the best possible value for the importance weight again the *LOWA* operator has to be applied.

Here the final outcome is the importance weights of the parameters.

Step 5: The performance value and the importance weight of the parameter are aggregated using the merging method as stated in Subsection 6.6.1.

7 Modeling with the Fuzzy Linguistic 2-tuple Approach with Unbalanced Linguistic Terms Set

7.1 AN OVERVIEW

In chapter 6, basically we have discussed several factor-based models developed upon uniformly and symmetrically distributed linguistic terms sets. As mentioned in previous chapters, there are situations when the experts' opinions require setting up the linguistic scale. However, it is not realistic to accept that all decision makers should agree on the same membership function associated to linguistic terms (Dong, Li, and Herrera, 2016; Herrera, Herrera-Viedma, and Martinez, 2008). This fact may lead to creating unbalanced terms in a particular linguistic scale. Figure 7.1 is a good interpretation for such an unbalanced scale.

Therefore it is worth investigating the way to develop models with qualitative/linguistic parameters where the performance values are evaluated based on an unbalanced linguistic terms set. To develop the models we use the fuzzy 2-tuple unbalanced representation algorithm introduced by Herrera et al. (Herrera, Herrera-Viedma, and Martinez, 2008).

Here we present two factor-based models where in Model I, the parameters are considered to be equally important and Model II associates with parameters which are unequally important.

In the next section, we present the systematic approach to build up the model.

7.2 SEMANTICS OF UNBALANCED LINGUISTIC TERMS SET

In this task, our aim is to assign the semantics of the labels in the unbalanced linguistic terms set. Unlike the uniform and symmetric linguistic scale, one should assign semantics in the unbalanced scale following the process as given in Figure 7.2. Below we explain the steps of the process in detail:

(I) Inputs
One may note that, to execute this process, three inputs are needed.

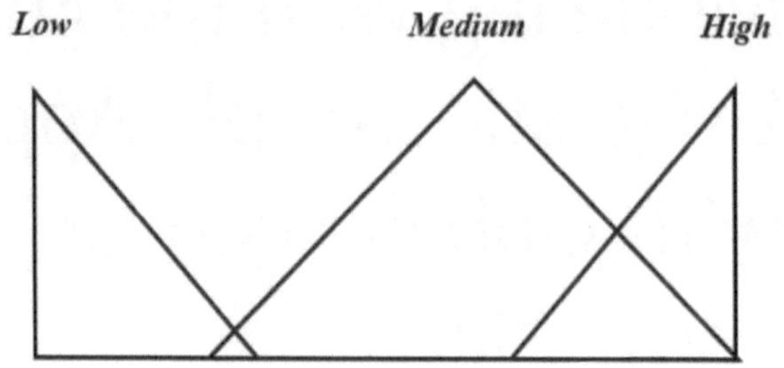

Figure 7.1 A sample illustration of an unbalanced linguistic terms set.

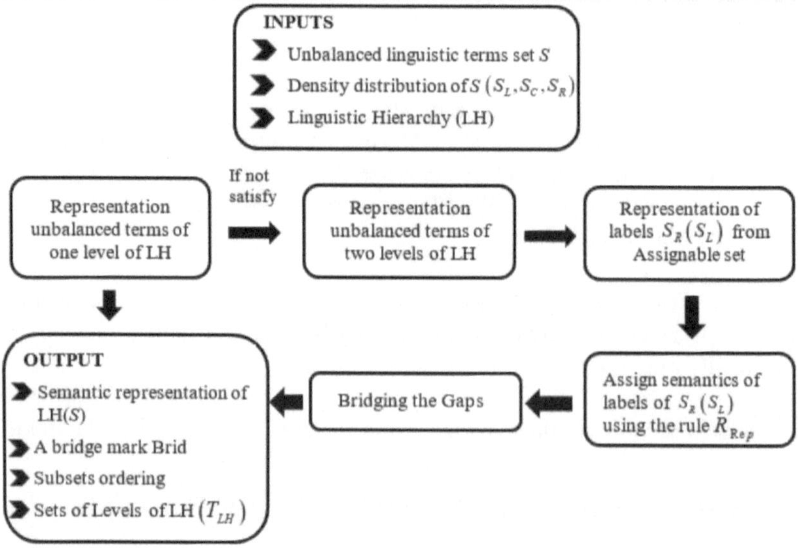

Figure 7.2 Semantic representation of an unbalanced linguistic terms set.

- As such, the first thing is to obtain the views from the experts on how the labels should be distributed within the given scale. Figure 7.1 is a example for an unbalanced linguistic scale.
- After obtaining the unbalanced linguistic terms set S, it has to be divided into three subsets such as S_L, S_C, and S_R where $S = S_L \cup S_C \cup S_R$. The subset S_C consists of only one label, i.e. central/middle label. The subset S_L consists of labels which are lesser than the central label where S_R consists of labels higher than the central label.
  Now the density distribution of S may be expressed as
  $$\{(Cars_{S_L}, density_{S_L}), \#(S_C), (Cars_{S_R}, density_{S_R})\} \tag{7.1}$$
  where $Cars_{S_L}$, $Cars_{S_R}$ denote the cardinality of S_R and S_L respectively. The $density_{S_L}$ and $density_{S_R}$ are symbolic variables in the set $Middle, Extreme$ which represent the higher granularity of the sets S_R/S_L whether it is concentrated near the central label or the Maximum(Minimum) label.
- The third input is the Linguistic Hierarchy (LH).
  LH consists of levels where these levels indicate a linguistic terms set with different granularity placed in a hierarchical manner (Herrera, Herrera-Viedma, and Martinez, 2008; Abrerizo, et al., 2017; Cai, Gong, and Yu, 2017; Yu, et al., 2017). Each level is denoted by $l(t, n(t))$ where t denotes the number that indicates the level of hierarchy and $n(t)$ is the granularity of the linguistic term set of t.
  LH can be represented as the union of all the levels $LH = \cup_t l(t, n(t))$.
  Linguistic labels in each level of LH are denoted as $S^{n(t)} = \left\{s_0^{n(t)}, \ldots, s_{n(t)-1}^{n(t)}\right\}$.
  Below we reproduce the transformation function which transforms a linguistic label in level t to a label in the consecutive level $t + c$, with $c \in \{-1, 1\}$, as defined in (Herrera, Herrera-Viedma, and Martinez, 2008).
  $$TF_{t+c}^{t}\left(s_i^{n(t)}, \alpha^{n(t)}\right) = \Delta_{t+c}\left(\frac{\Delta_t^{-1}\left(s_i^{n(t)}, \alpha^{n(t)}\right) \cdot (n(t+c) - 1)}{n(t) - 1}\right). \tag{7.2}$$
  To set up an LH the following conditions must be satisfied:

  i. Retain all former mid-points of the membership functions of each linguistic term from one level to the following one.
  ii. The linguistic term set of the level $t+1$, $S^{n(t+1)}$, is obtained from $S^{n(t)}$ as
  $$l(t, n(t)) \rightarrow l(t+1, 2 \cdot n(t) - 1). \tag{7.3}$$
  For example, if $n(t)$ is 3 in level 1 then $n(t+1)$ for level 2 is obtained as 5 using Eq. (7.2). Figure 7.3 illustrates a sample LH set-up using 3 levels.

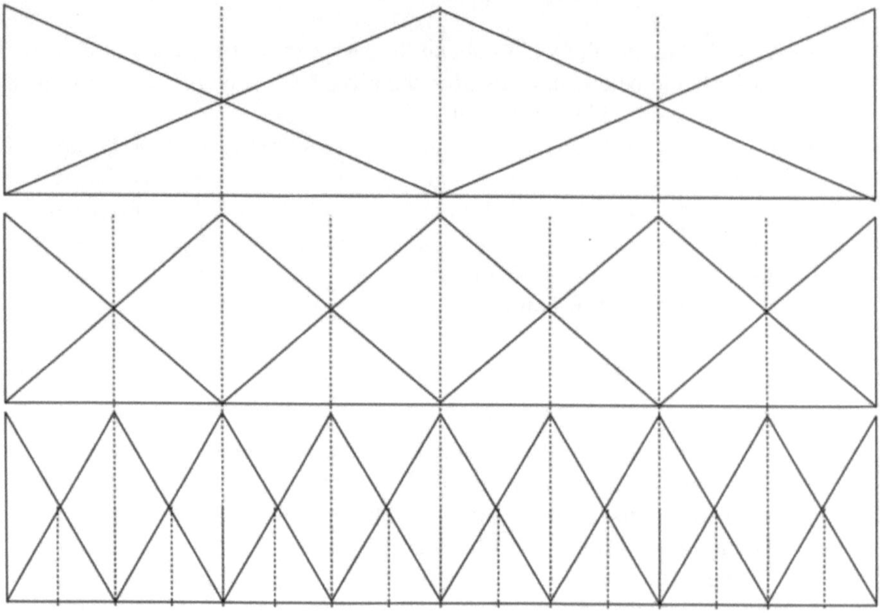

Figure 7.3 Linguistic hierarchy of 3, 5, and 9 labels. (Modified from *Linguistic Multi-criteria Decision-making Model with Output Variable Expressive Richness,* 83, Cid-López, A., Hornos, M.J., Carrasco, R.A., Herrera-Viedma, E. and Chiclana. 350-362, Copyright (2017), with permission from Elsevier.)

(II) Representation of unbalanced terms of one level of *LH*

Here our aim is to represent the labels of set $S_R\,(S_L)$ from the labels in the levels of *LH*.

First of all, we need to check that the following condition is satisfied (Herrera, Herrera-Viedma, and Martinez, 2008):

$$\exists t \in LH, \frac{n(t)-1}{2} = Car_{S_R}, \text{or}, \frac{n(t)-1}{2} = Car_{S_L}. \quad (7.4)$$

The above condition admits there exists a level t in LH whose cardinality of the subset is same as the cardinality of the subsets $S_R\,(S_L)$.

If the condition (Eq.(7.4)) is satisfied, the labels of the $S_R\,(S_L)$ may be represented as follows (Herrera, Herrera-Viedma, and Martinez, 2008):

- The labels of $S_R\,(S_L)$ assigned from $S_R^{n(t)}\left(S_L^{n(t)}\right)$, i.e., $S_R \leftarrow S_R^{n(t)}\left(S_L \leftarrow S_L^{n(t)}\right)$.

- Assign labels to central subset $S_C = \{s_C\}$.

  The downside of the central label is assigned by the downside of the central label $s_C^{n(t)} \in S^{n(t)}$, i.e., $\underline{s_C} \leftarrow \underline{s_C^{n(t)}}$, when dealing with the set S_R.

  The upside of the central label is assigned by the upside of the label $\overline{s_C} \leftarrow \overline{s_C^{n(t)}}$ when dealing with the set S_L.

(III) Representation of unbalanced terms of two levels of *LH*

In this case, one may need to consider two levels of *LH* to represent the labels of $S_R\,(S_L)$ when the condition (Eq.(7.4)) is not satisfied (Herrera, Herrera-Viedma, and Martinez, 2008).

For this representation, the density distribution of S, i.e. (Herrera, Herrera-Viedma, and Martinez, 2008),
$\{(Car_{S_L}, density_{S_L}), \#(S_C), (Car_{S_R}, density_{S_R})\}$ is considered.

The subsequent steps explain the mechanism for this representation.

(IV) Representation of labels $S_R\,(S_L)$ from assignable set

Assignable sets $AS_R^{n(t)}\left(AS_L^{n(t)}\right)$ and $AS_R^{n(t+1)}\left(AS_L^{n(t+1)}\right)$ are the right(left) lateral set which represents the terms of $S_R\,(S_L)$ (Herrera, Herrera-Viedma, and Martinez, 2008).

The sets $AS_R^{n(t)}$ and $AS_R^{n(t+1)}$ consist of

$AS_R^{n(t)} = S_R^{n(t)} = \left\{s_{((n(t)-1)/2)+1}^{n(t)}, \ldots, s_{n(t)-1}^{n(t)}\right\}$, and

$AS_R^{n(t+1)} = S_R^{n(t+1)} = \left\{s_{((n(t+1)-1)/2)+1}^{n(t+1)}, \ldots, s_{n(t+1)}^{n(t+1)}\right\}$.

One may note that the following procedures are symmetrically correspondent to the set S_L.

First, the set S_R is divided into two subsets S_{RC}, and S_{RE}, such that $S_R = S_{RC} \cup S_{RE}$. Here the set S_{RC} consists of the labels close to S_C and S_{RE} consists of labels close to the Maximum label in S.

Below we discuss the rule to represent the labels of S_R from $AS_R^{n(t)}$ and $AS_R^{n(t+1)}$ (Herrera, Herrera-Viedma, and Martinez, 2008):

If $density_{S_R}$="extreme"
Represent S_{RE} on $AS_R^{n(t+1)}$, i.e., $S_{RE} \subset AS_R^{n(t+1)}$
Represent S_{RC} on $AS_R^{n(t)}$, i.e., $S_{RC} \subset AS_R^{n(t)}$
If $density_{S_R}$="middle"
Represent S_{RE} on $AS_R^{n(t)}$, i.e., $S_{RE} \subset AS_R^{n(t)}$

Represent S_{RC} on $AS_R^{n(t+1)}$, i.e., $S_{RC} \subset AS_R^{n(t+1)}$

(V) Assign semantics of labels of $S_R\,(S_L)$ using the rule R_{Rep} (Herrera, Herrera-Viedma, and Martinez, 2008)

In this task, the semantics of the labels $S_R\,(S_L)$ are assigned by sets $AS_R^{n(t)}$ and $AS_R^{n(t+1)}$ using the rule R_{Rep}, i.e. (Herrera, Herrera-Viedma, and Martinez, 2008),

when a label $s_i^R \in S_R$ is represented by label $s_k^{n(t+1)} \in AS_R^{n(t+1)}$ $k = 2 \cdot j$ or $k = 2j - 1$ then eliminate $s_k^{n(t+1)}$ from $AS_R^{n(t+1)}$ and eliminate the label $s_j^{n(t)} \in AS_R^{n(t)}$.

We denote $Nlab_t$ as the number of labels in $AS_R^{n(t)}$ to represent labels of S_{RE} and the formula is given as

$Nlab_t = ((n(t+1) - 1)/2) - Car_{S_R}$.

In the following, we present the procedure of semantic representation by using R_{Rep} and representation functions (Herrera, Herrera-Viedma, and Martinez, 2008):

Here the representation functions are categorized in accordance with the $S_R\,(S_L)$ and the levels t and $t+1$.

- The function $Assign_{t+1}^R\,(density)$ is defined for the representation of the set S_R in level $t+1$ of LH.
  It produces the representations of unbalanced linguistic terms in S_R by using the $AS_R^{n(t+1)}$ of level $t+1$. Also this function depends on the symbolic value density, i.e., middle or extreme.

Case I: When density=middle

By following R_{Rep} and beginning with the label following the middle label $s_{C+1}^{n(t+1)}$, the $Nlab_{t+1}$ labels in S_{RE} are represented by the smallest labels contained in AS_R^{n+1}.

Case II: When density=extreme

By following R_{Rep} and beginning with the highest label $S^{n(t+1)}_{n(t+1)-1}$, the $Nlab_{t+1}$ labels in S_{RE} represent the largest labels contained in $AS^{n(t+1)}_R$.

- The function $Assign^R_t(density)$ is defined for the representation of S_R in level t of LH.

  It produces the representations of unbalanced linguistic terms in S_R by using the $AS^{n(t)}_R$ of level t.

  Case I: When density=middle

  $Nlab_t$ labels in $S_{RE} \subset S_R$ represent the highest labels contained in $AS^{n(t)}_R$ and begin with $s^{n(t+1)}_{C+1+\delta}$ where $\delta = round\,((lab_{t+1}/2))$

  Case II: When density=extreme

  Here $Nlab_t$ labels in $S_{RC} \subset S_R$ represent the smallest labels contained in $AS^{n(t)}_R$ and begin with $S^{n(t)}_{n(t)-1-\delta}$.

- The function $Assign^L_{t+1}(density)$ produces the representation of S_L in the level $t+1$ of LH.
  Using this function with the set $AS^{n(t+1)}_L$ of level $t+1$ produces the representation of unbalanced linguistic terms in the set S_R. Depending on the value of the symbolic parameter $density \in \{middle, extreme\}$, two cases can be presented as follows:

  Case I: When density=middle
  By following R_{Rep} and beginning with the previous to the middle label $s^{n(t+1)}_{C-1}$, the $Nlab_{t+1}$ labels in the set $S_{LC} \subset S_L$ represent the largest labels in $AS^{n(t+1)}_L$.

  Case II: When density=extreme
  By following R_{Rep} and beginning with the smallest label $s^{n(t+1)}_0$, the $Nlab_{t+1}$ labels in $S_{LE} \subset S_L$ represent the smallest labels contained in $AS^{n(t+1)}_L$.

- The function $Assign^L_t(density)$ generates the representation of S_L in the level t of LH along with the set $AS^{n(t)}_L$ of level t.

  Case I: When density=middle
  $Nlab_t$ in $S_{LE} \subset S_L$ represents the smallest labels in $AS^{n(t)}_L$ and starts with $s^{n(t)}_{C-1-\delta}$ where $\delta = round\,((lab_{t+1})/2)$.

  Case II: When density=extreme
  $Nlab_t$ in $S_{LC} \subset S_L$ represents the highest labels in $AS^{n(t)}_L$ and starts with $s^{n(t)}_\delta$.

(VI) Bridging the Gaps

Here we consider a label in S which shows jumps between the levels t and $t+1$, denoted as s_J. These jumps need to be bridged using the following procedure (Herrera, Herrera-Viedma, and Martinez, 2008):

when $density_{S_R}$=extreme

$$\overline{s_J} \leftarrow \overline{s_i^{n(t)}}, \underline{s_J} \leftarrow \underline{s_k^{n(t+1)}}, k = 2 * i$$

when $density_{S_R}$=middle

$$\underline{s_J} \leftarrow \underline{s_i^{n(t)}}, \overline{s_J} \leftarrow \overline{s_k^{n(t+1)}}, k = 2 * i$$

(VII) Output

As depicted in Figure 7.2, there are four outputs in this representation. Each output is briefly explained as follows (Herrera, Herrera-Viedma, and Martinez, 2008):

(i) Semantic representation of LH(S)

The representation of the unbalanced linguistic terms set denoted as $S = \{s_i, i = 0, ..., g\}$ by means of LH is defined as $LH(S) = \left\{s_I^{G(i)}, i = 0, ..., g\right\}$, $\forall s_i \in \exists l(t, n(t) \in LH$ that contains a label $s_k^{n(t)} \in S_{n(t)}$ such that $Index(i) = k$ and $Granual(i) = n(t)$.
Here $Index(i)$ denotes the index of the label represented in LH, where $Granual(i) = n(t)$ denotes the granularity of levels in LH.

(ii) A bridge mark Brid
The function $Brid : S \rightarrow \{True, False\}$ is defined for the labels s_J.

(iii) Subset ordering
The subsets of the set S are arranged in increasing order.

(iv) Sets of levels of LH (T_{LH})
$t_{LE}, t_{LC}, t_{RC}, t_{RE}$ represent the levels of LH corresponding to the sets $S_{LE}, S_{LC}, S_{RC}, S_{RE}$.

In the following, we discuss the method to develop a factor-based model where the performance values of the parameters are evaluated from the unbalanced linguistic terms set.

7.3 MODEL BASED ON UNBALANCED LINGUISTIC TERMS SET

The problems which consider qualitative/linguistic factors as parameters can apply this model for decision making. Figure 7.4 depicts the structure of the model.

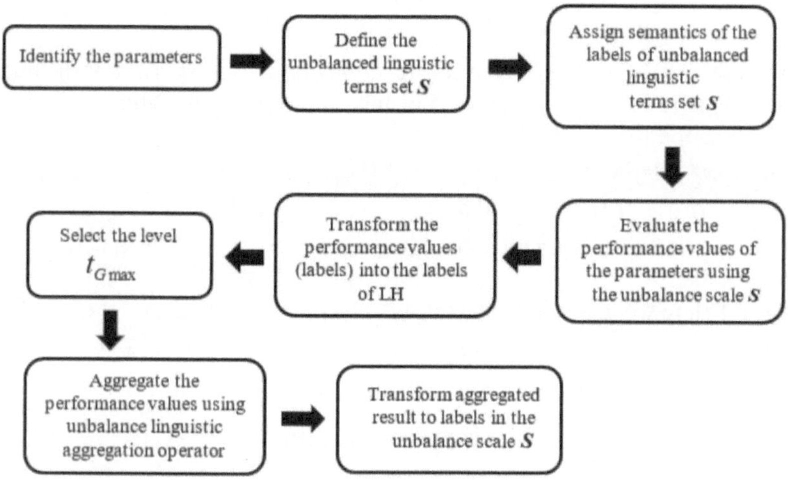

Figure 7.4 Structure of the model-unbalanced scale.

Below we explain the steps of the model structure in detail.

Steps 1 to 3: After identifying the parameters, the unbalanced terms set has to be set up by considering the experts' views. The method discussed in Section 7.2 should follow to assign the semantics of labels in the terms set.

Step 4: Here the performance values of the parameters are evaluated using the unbalanced linguistic scale. These values should be in the form of 2-tuples, i.e. (s_i, α).

Step 5: In this step, the performance values which are in the form of 2-tuples, i.e. $s_i \in S$, need to be transformed into linguistic terms $LHs_k^{n(t)} \in LH = \cup_t l(t, n(t))$, and viceversa.

Below we reproduced two transformation functions as defined in (Herrera, Herrera-Viedma, and Martinez, 2008):

(1) The function which is denoted as L_H transforms the unbalanced 2-tuple $s_i \in S$ into the corresponding linguistic 2-tuple in $LHs_k^{n(t)}$, $s_k^{n(t)} \in LH$.
The function for L_H is:
$L_H : (S \times [-0.5, 0.5)) \rightarrow (LH \times [-0.5, 0.5))$
such that $\forall (s_i, \alpha_i) \in (S \times [-0.5, 0.5))$

$$\exists L_H(s_i,\alpha_i)=\left(S_{I(i)}^{G(i)},\alpha_i\right), S_{I(i)}^{G(i)}\in LH.$$

(2) The function L_H^{-1} acts as the inverse function of L_H; i.e. it transforms the linguistic 2-tuple in LH to its corresponding unbalanced linguistic 2-tuple $((s_i,\alpha)$. The function is:
$L_H^{-1}:(LH\times[-0.5,0.5))\rightarrow(S\times[-0.5,0.5))$,
$\forall\left(s_k^{n(t)},\alpha_k\right)\in(LH\times[-0.5,0.5))\,|s_k^{n(t)}\in S_{n(t)}, t\in LH.$
This function executes with several cases and these are presented below:

Case 1 Here we focus on the fact that the function L_H^{-1} generates the unbalanced term directly from the $s_k^{n(t)}\in LH(S)$, i.e. $L_H^{-1}\left(s_k^{n(t)},\alpha_k\right)=(s_i,\lambda)$.
In order to proceed the following condition should be satisfied:

$$\exists s_i\in S|G(i)=n(t)\text{and}I(i)=k. \tag{7.5}$$

One may note the value λ is unknown. To determine the value, Eq. (7.6) is defined as in (Herrera, Herrera-Viedma, and Martinez, 2008).

$$\lambda=\left(\frac{\Delta_t^{-1}\left(s_k^{n(t)}\right)\cdot(n(t+1)-1)}{n(t)-1}\right)-\text{round}\left(\frac{\Delta_t^{-1}\left(s_k^{n(t)}\right)\cdot(n(t+1)-1)}{n(t)-1}\right). \tag{7.6}$$

Equation (7.6) executes with two cases as follows (Herrera, Herrera-Viedma, and Martinez, 2008):

Case (1.1)
If Brid(s_i)=False
$\lambda=\alpha_k$, with $\mathscr{L}\mathscr{H}^{-1}\left(s_k^{n(t)},\alpha_k\right)=(s_i,\alpha_k)$.
Case (1.2) If Brid(s_i)=True
In this case, the semantics of label s_i are represented with two levels of LH. This aspect has been categorized into three parts as follows:

A. If $s_i\in\mathscr{S}_{RE}$ or $s_i\in\mathscr{S}_{LC}$ then $\overline{s_i}$ is defined from $TF_{t+1}^{t}\left(s_k^{n(t)},0\right)=s_{2k}^{n(t+1)}$ while s_i is defined from $s_k^{n(t)}$. This will lead to two possibilities presented as follows:

(a) If $s_i\in\mathscr{S}_{RE}$ or $s_i\in\mathscr{S}_{LC}$ then $\overline{s_i}$ is defined from $TF_{t+1}^{t}\left(s_k^{n(t)},0\right)=s_{2k}^{n(t+1)}$ while s_i is defined from $s_k^{n(t)}$. This will lead to two possibilities presented as follows:

(i) If α_k represents a symbolic translation on $\overline{s_k^{n(t)}}$(upside part of the membership function) then
$\alpha_k<0\Rightarrow\lambda\in[-0.5,0)$. $\overline{s_k^{n(t)}}$ belongs to the level t, λ is computed using Eq. (7.6).

(ii) If α_k represents a symbolic translation on $\underline{s_k^{n(t)}}$ (downside part of the membership function) $\alpha_k > 0 \Rightarrow \lambda \in [0, 0.5)$ with $\lambda = \alpha_k$.

(b) If $s_i \in \mathscr{S}_{RC}$ or $s_i \in \mathscr{S}_{LE}$, then $\overline{s_i}$ is defined from $\overline{s_k^{n(t)}}$ while $\underline{s_i}$ is defined from $TF_{t+1}^{t}\left(s_k^{n(t),0}\right) = s_{2k}^{n(t+1)}$. Therefore the two possibilities are

(i) If αk represents a symbolic translation on $\overline{s_k^{n(t)}}$ then $\alpha_k < 0 \Rightarrow \lambda \in [-0.5, 0)$ where $\lambda = \alpha_k$.

(ii) If α_k represents a symbolic translation on $\underline{s_k^{n(t)}}$, then $\alpha_k > 0 \Rightarrow \lambda \in [0, 0.5)$ and λ is computed using Eq. (7.6).

(c) If $s_i \in \mathscr{S}_{\mathscr{C}}$ then three possibilities can be found by considering the levels of LH used to represent the semantics of s_i, t_{LC}, and t_{RC}.

(i) If $t_{LC} = t_{RC}$, then $\mathscr{L}\mathscr{H}^{-1}\left(s_k^{n(t)}, \alpha_k\right) = (s_i, \alpha_k)$.

(ii) If $t_{LC} > t_{RC}$, then

$$\mathscr{L}\mathscr{H}^{-1}\left(s_k^{n(t)}, \alpha_k\right) = (s_i, \lambda), \begin{cases} \lambda = \alpha_k & \text{if } \alpha_k > 0 \\ \lambda = \text{Eq.}(7.6) & \text{if } \alpha_k < 0. \end{cases}$$

(iii) If $t_{LC} < t_{RC}$, then

$$\mathscr{L}\mathscr{H}^{-1}\left(s_k^{n(t)}, \alpha_k\right) = (s_i, \lambda), \begin{cases} \lambda = \alpha_k & \text{if } \alpha_k < 0 \\ \lambda = \text{Eq.}(7.6) & \text{if } \alpha_k > 0. \end{cases}$$

Case (2) If Eq. (7.5) is not satisfied, then $\mathscr{L}\mathscr{H}^{-1}\left(s_k^{n(t)}, \alpha_k\right) = \mathscr{L}\mathscr{H}^{-1}\left(TF_{t'}^{t}\left(s_k^{n(t)}, \alpha_k\right)\right)$ with $t' \in \{t_{LE}, t_{LC}, t_{RC}, t_{RE}\}$ being a level such that if $TF_{t'}^{t}\left(s_k^{n(t)}, \alpha_k\right) = \left(s_{k'}^{n(t')}, \alpha_{k'}\right)$. Then $\exists s_i \in \mathscr{S} | G(i) = n(t')$ and $I(i) = k'$.

Select the level t_{Gmax} Here the level t_{Gmax} which represents the highest granularity from the set $\{t_{LE}, t_{LC}, t_{RC}, t_{RE}\}$ has to be chosen.

Step 6 Aggregate the performance values using the unbalanced linguistic aggregation operator.

To aggregate the unbalanced linguistic values in the form of 2-tuples, these should be defined in the domain $S^{n(t_{Gmax})}$ (Herrera, Herrera-Viedma, and Martinez, 2008).

In order to place the unbalanced linguistic 2-tuples in the same domain as $S^{n(t_{Gmax})}$, the following transformation function is reproduced as stated in (Herrera, Herrera-Viedma, and Martinez, 2008).

Consider the unbalanced 2-tuples t (s_i, α_i) $(s_i \in S)$ and $\left(s_k^{n(t')}, \alpha_k\right) = LH(s_i, \alpha_i)$. $\left(s_k^{n(t')}, \alpha_k\right)$ represents the corresponding 2-tuple of t (s_i, α_i) in the level of $t' \in \{t_{LE}, t_{LC}, t_{RC}, t_{RE}\}$.

The following cases support transforming the $\left(s_k^{n(t')}, \alpha_k\right)$ in level t' into the 2-tuple in level t_{Gmax}:

Case 1 Case (1) If Brid(s_i)=false

The semantic of s_i can be represented only with one label in LH, and therefore, $\mathcal{TF}_{t_{Gmax}}^{t'}\left(s_k^{n(t')}, \alpha_k\right) = TF_{t_{Gmax}}^{t'}\left(s_k^{n(t')}, \alpha_k\right), \forall t'$.

Case 2 If Brid(s_i)=true then the semantic representation of s_i is associated with two labels in LH and the definition of $\mathcal{TF}_{t_{HGLS}}^{t'}$ depends on the localization of s_i in $\mathcal{S}$ which can be divided into three cases as follows:

Case 2.1 If $s_i \in \mathcal{S}_{RE}$ or $s_i \in \mathcal{S}_{LC}$ ($t' = t_{RE}$ and $t'+1 = t_{RC}$ or $t' = t_{LC}$ and $t'+1 = t_{LE}$)

In this case $\overline{s_i}$ is defined from $\mathcal{TF}_{t'+1}^{t'}\left(s_k^{n(t')}, 0\right)$ and $\underline{s_i}$ is defined from $s_k^{n(t')}$. For this representation one may follow the two possibilities as given below:

(i) If $\alpha_k < 0$ (upside of $s_k^{n(t')}$)
$\mathcal{TF}_{t_{Gmax}}^{t'}\left(s_k^{n(t')}, \alpha_k\right) = \mathcal{TF}_{t_{Gmax}}^{t'+1}\left(s_{2k}^{n(t'+1)}, \alpha_k\right)$.

(ii) If $\alpha_k \geq 0$ (downside of $s_k^{n(t')}$)
$\mathcal{TF}_{t_{Gmax}}^{t'}\left(s_k^{n(t')}, \alpha_k\right) = \mathcal{TF}_{t_{Gmax}}^{t'}\left(s_k^{n(t')}, \alpha_k\right)$.

Case 2.2 If $s_i \in \mathcal{S}_{RC}$ or $s_i \in \mathcal{S}_{LE}$ ($t' = t_{RC}$ and $t'+1 = t_{RE}$ or $t' = t_{LE}$ and $t'+1 = t_{LC}$)
The $\overline{s_i}$ is defined from $\overline{s_k^{n(t')}}$ and $\underline{s_i}$ is defined from $TF_{t+1}^{t}\left(s_k^{n(t)}, 0\right)$. For this case, one may need to follow up two possibilities as follows:

(i) If $\alpha_k \leq 0$ (upside of $s_k^{n(t')}$)

$\mathcal{TF}_{t_{Gmax}}^{t'}\left(s_k^{n(t')}, \alpha_k\right) = TF_{t_{Gmax}}^{t'}\left(s_k^{n(t')}, \alpha_k\right)$.

(ii) If $\alpha_k > 0$ (downside of $s_k^{n}(t')$) then
$\mathcal{TF}_{t_{Gmax}}^{t'}\left(s_k^{n(t')}, \alpha_k\right) = TF_{t_{Gmax}}^{t'+1}\left(s_{2k}^{n(t'+1)}, \alpha_k\right)$.

Case 2.3 If $s_i \in \mathscr{S}_C$

Here the levels t_{LC} and t_{RC} are considered with the following three possibilities:

(i) If $t_{LC} = t_{RC}$

$$\mathscr{T}\mathscr{F}^{t'}_{t_{Gmax}}\left(s_k^{n(t')}, \alpha_k\right) = TF^{t'}_{t_{Gmax}}\left(s_k^{n(t')}, \alpha_k\right).$$

(ii) If $t_{LC} = t_{RC}$ $\mathscr{T}\mathscr{F}^{t'}_{t_{Gmax}}\left(s_k^{n(t')}, \alpha_k\right) = \begin{cases} TF^{t'}_{t_{Gmax}}\left(s_k^{n(t')}, \alpha_k\right) & \text{if } \alpha_k \geq 0 \\ TF^{t'+1}_{t_{Gmax}}\left(s_{2k}^{n(t'+1)}, \alpha_k\right) & \text{if } \alpha_k < 0. \end{cases}$

(iii) If $t_{LC} < t_{RC}$

$$\mathscr{T}\mathscr{F}^{t'}_{t_{Gmax}}\left(s_k^{n(t')}, \alpha_k\right) = \begin{cases} TF^{t'}_{t_{Gmax}}\left(s_k^{n(t')}, \alpha_k\right) & \text{if } \alpha_k \leq 0 \\ TF^{t'+1}_{t_{Gmax}}\left(s_{2k}^{n(t'+1)}, \alpha_k\right) & \text{if } \alpha_k > 0. \end{cases}$$

After placing the linguistic 2-tuples in the same domain as $S^{nt_{Gmax}}$, these can be aggregated using a unbalanced linguistic aggregation operator.

Below we reproduced the generic aggregation operator of unbalanced linguistic information as defined in (Herrera, Herrera-Viedma, and Martinez, 2008).

Definition 7.1. (Herrera, Herrera-Viedma, and Martinez, 2008)

Let $A = \{(a_1, \alpha_1), ..., (a_p, \alpha_p), a_i \in [-0.5, 0.5)\}$ be a set of unbalanced linguistic assessment to be aggregated. Then a generic aggregation operator of unbalanced linguistic information $\Lambda^F : (\mathscr{S} \times [-0.5, 0.5))^p \rightarrow \mathscr{S} \times [-0.5, 0.5)$ is defined according to the following expression: $\Lambda^F [(a_1, \alpha_1), ..., (a_p, \alpha_p)] = \mathscr{L}\mathscr{H}^{-1}\left(s_k^{n(t_{Gmax})}, \lambda\right)$ with $\left(s_k^{n(t_{Gmax})}, \lambda\right)$ being the linguistic 2-tuple obtained as

$\left(s_k^{n(t_{Gmax})}, \lambda\right) = Agg\left(\mathscr{T}\mathscr{F}^{t'}_{t_{Gmax}}(\mathscr{L}\mathscr{H}(a_1, \alpha_1)), ..., \mathscr{T}\mathscr{F}^{t'}_{t_{Gmax}}(\mathscr{L}\mathscr{H}(a_p, \alpha_p))\right)$,

$t', t_{Gmax} = max\{t_{LE}, t_{LC}, t_{RC}, t_{RE}\}$, and Agg any aggregation operator of linguistic 2-tuples.

Step 7 Transform aggregated result to labels in the unbalanced scale S.

This is the final step, which transforms the aggregated result into a label in the unbalanced linguistic terms set S. This can done by using the function LH^{-1} transformation function.

8 Evaluation of Invasion Risk with Quantitative Parameters

8.1 QUANTITATIVE PARAMETERS/FACTORS IN INVASIVENESS

In this section, our main aim is to elaborate upon the procedure to develop models based on the model structures discussed in Chapter 4 to evaluate the risk of Invasive Alien Plant Species (IAPS). The model structures mentioned in Chapter 4 have been developed to work with quantitative parameters. Therefore, the first task is to identify the biological traits/parameters in the quantitative form which directly/indirectly affect the invasiveness of IAPS. After performing a literature survey and having discussions with plant science experts, four parameters in quantitative form which are related to the dispersal trait have been identified (Marambe, et al., 2001; Kotagama and Bambardeniya, 2006; Chenje and Mohamed-Katerere, 2006; Natural Heritage, 2013; Ranwala, et al., 2012). These parameters are:

- Number of seeds in a fruit (NSF)
- Annual Seed Rain per m^2 (ASR)
- Seeds Viability (in months) (VIA)
- Long Distance Dispersal Strength (LDD)

Based on the assumption that these four parameters describe the dispersal risk of IAPS, we apply the model presented in Figure 4.5 to build up dispersal risk models. In the following section, we discuss the steps to obtain the specified dispersal risk models in detail.

8.2 EVALUATION OF DISPERSAL RISK MODELS

One may note that the model presented in Figure 4.5 illustrates main four steps such as data collection, fuzzification of parameters, parameter aggregation, checking the reliability of output, and the final output. Now let us discuss each step relating to construction of models which produce the final outcome as the dispersal risk of IAPS.

- Step 1. Data collection

Data of 22 IAPS relating to four parameters have been gathered using trusted sources and the National Risk Assessment (NRA) protocol established invasive specialist group attached to the Ministry of Environment, Sri Lanka (Ranwala, et al., 2012). In addition, invasion risk scores/NRA scores of these IAPS which are obtained from the NRA procedure using the check-list prepared and accepted by the Ministry of Environment and Renewable Resources, Sri Lanka after broad stakeholder participation and discussion have also been gathered.

Next we discuss the fuzzification of four parameters.

- Step 2. Fuzzy sets and membership functions of dispersal risk parameters
  In this task, our aim is to convert the above mentioned four parameters into fuzzy parameters. For that the fuzzy set for each parameter has to be defined. First the domain of each parameter needs to be defined. It is not possible to define the boundaries directly by only looking at the performance values of parameters, due to the imprecise and ill-defined domains. Views of the group of plant science experts have been taken to set up the lower and upper limits of the domains together with assumptions in determining the limit points as stated below (Peiris, et al., 2017, 326-339, Peiris, et al., 2017, 6-14):

  (i) lower limit is the least possible value which shows the slightest effect on the dispersal risk of plant species.
  (ii) upper limit is the highest value which shows the utmost effect on the dispersal risk of plant species. The upper limit is chosen as an unrealistic extreme value to be compatible with any invasive plant other than in the database.
  (iii) dispersal of plant species increases when the values of parameters increase from lower to upper limits.

  Now we define fuzzy sets $\tilde{S}_{NSF}$, $\tilde{S}_{ASR}$, $\tilde{S}_{VIA}$, and $\tilde{S}_{LDD}$ for the parameters number of seeds per fruit (NSF), annual seed rain per m^2 (ASR), viability of seeds in months (VIA), long distance dispersal strength (LDD), respectively as follows:

$$\tilde{S}_{NSF} = \{(x, \mu_{NSF}(x)) \,|x \varepsilon R, \mu_{NSF}(x) \, \varepsilon \, [0,1]\}. \tag{8.1}$$

$$\tilde{S}_{ASR} = \{(x, \mu_{ASR}(x)) \,|x \varepsilon R, \mu_{ASR}(x) \, \varepsilon \, [0,1]\}. \tag{8.2}$$

$$\tilde{S}_{VIA} = \{(x, \mu_{VIA}(x)) \,|x \varepsilon R, \mu_{VIA}(x) \, \varepsilon \, [0,1]\}. \tag{8.3}$$

$$\tilde{S}_{LDD} = \{(x, \mu_{LDD}(x)) \,|x \varepsilon R, \mu_{LDD}(x) \, \varepsilon \, [0,1]\}. \tag{8.4}$$

  Here $\mu_{NSF}(x)$,$\mu_{ASR}(x)$,$\mu_{VIA}(x)$, and $\mu_{LDD}(x)$ are unknown membership functions of the parameters, NSF, ASR, VIA, LDD, respectively.

  Now the computational challenge is to set up the suitable membership function for each fuzzy set in order to show the degree of relevance at each point in the domain appropriately. For this task, considering the assumptions mentioned above and the experts' views, the Z-shape function has been chosen as the membership function for each fuzzy set initially. The membership function for each fuzzy set is defined as follows:

$$\mu_{NSF}(x) = \begin{cases} 1 & \text{for } x < 1 \\ 1 - 2\left[(x-1)/1000\right]^2 & \text{for } 1 \le x \le 501 \\ 2\left[(x-1001)/1000\right]^2 & \text{for } 501 < x \le 1001 \\ 0 & \text{for } x > 1001. \end{cases} \tag{8.5}$$

$$\mu_{ASR}(x) = \begin{cases} 1 & \text{for } x \le 0 \\ 1 - 2\left(\dfrac{x-0}{10^7}\right)^2 & \text{for } 0 \le x \le \dfrac{10^7}{2} \\ 2\dfrac{(10^7 - x)^2}{(10^7)^2} & \text{for } \dfrac{10^7}{2} \le x \le 10^7 \\ 0 & \text{for } x \ge 10^7. \end{cases} \tag{8.6}$$

$$\mu_{VIA}(x) = \begin{cases} 1 & \text{for } x \le 3 \\ 1 - 2\left[(x-3)/1197\right]^2 & \text{for } 3 \le x \le \dfrac{1203}{2} \\ 2\left[(1200-x)/1197\right]^2 & \text{for } \dfrac{1203}{2} < x \le 1200 \\ 0 & \text{for } x \ge 1200. \end{cases} \tag{8.7}$$

$$\mu_{LDD}(x) = \begin{cases} 1 & \text{for } x \le 0 \\ 1 - 2\left[(x-0)/10\right]^2 & \text{for } 0 \le x \le 5 \\ 2\left[(10-x)/10\right]^2 & \text{for } 5 < x \le 10 \\ 0 & \text{for } x \ge 10. \end{cases} \tag{8.8}$$

One may note that the domain of the parameter *NSF* is defined from 1 to 1001 whereas *ASR* ranges from 0 to 10^7 and *VIA* is defined from 3 to 1200. Data for the *LDD* of plant species are represented in the form of a ten point scale which represents point 0 as the lowest dispersal strength while point 10 is the highest dispersal strength.

- Step 3. Aggregation of fuzzy parameters
  Here our aim is to aggregate the fuzzified parameters *NSF*, *ASR*, *VIA*, and *LDD* to determine the dispersal risk of IAPS. The term aggregation reveals the combined effect of the four parameters. Therefore, we use four parameterized intersection operators (t-norm) such as Hamacher, Yager, Dombi, and Dubois for this aggregation procedure. In the latter part of the chapter, four models which have been developed from each operator are explained.

- Step 4. Investigating the output reliability

After aggregating the four fuzzy parameters, one may need to investigate whether the output is compatible with the expected output. For this application, we have NRA scores for IAPS in the data set which are obtained from the experts in the plant science field. These NRA scores have been considered as the measure to check the reliability of the output.

For example, assume that the model output and NRA scores obtained for IAPS are compatible with each other insignificantly; the process needs to be redirected to Step 2. In that task, the membership function needs to be modified to eliminate the incompatibilities. After moving to Step 3, one may use the CON/DIL operators to weight the parameters to trigger the compatibility procedure. Likewise, until the model produces a reliable output, Steps 2 to 4 have to be performed back and forth.
Below, the models which have been developed based on the parameterized intersection operators are explained.

- Step 5. Final output

As we mentioned earlier, the final output of the model is dispersal risk. Here we need the dispersal risk in the form of risk levels similar to the linguistic labels depicted in Figure 4.4. As such, dispersal risk is segregated into seven risk levels: *Unlikely, Very Low, Low, Medium, High, Very High*. In Step 4, the output is in the form of numerical value, so using Eq. (4.3), the numerical value has been converted to a linguistic label.

In the following sections, we discuss the development of models based on the parameterized intersection operators.

8.3 MODEL I—HAMACHER *t*-NORM OPERATOR

This model is developed based on the parameterized $t-$norm Hamacher operator as defined in the Eq. (3.13). After applying Steps 2 to 4 in Section 8.2, the membership functions of *ASR*, *VIA*, *LDD* given in Eqs. (8.6) to (8.8) have been modified as in Eqs. (8.9) to (8.11). The graphical representations of these modified functions are depicted in Figures 8.1 to 8.3.

$$\mu_{ASR_{Ham}}(x) = \begin{cases} 2\dfrac{(10000-x)^2}{8\times 10^8}+0.75 & \text{for } 0 \le x < 10000 \\ 2\dfrac{(100000-x)^2}{5.4\times 10^{10}}+0.45 & \text{for } 10000 \le x < 100000 \\ 2\dfrac{(10\times 10^6-x)^2}{4.356\times 10^{14}} & \text{for } 100000 \le x \le 10^7 \\ 0. & \text{for } x > 10^7 \end{cases} \tag{8.9}$$

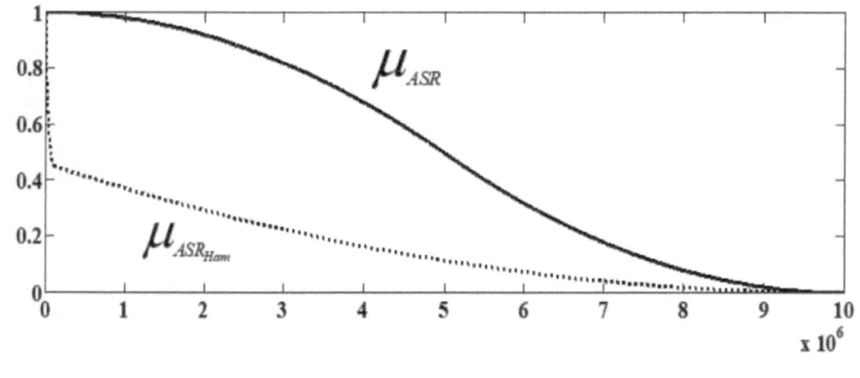

Figure 8.1 Original and modified membership functions of *ASR* in Model I-Hamacher. (Modified from H.O.W. Peiris, S. Chakraverty, S.S.N. Perera, S.S.N and S.M.W. Ranwala, "Modelling Dispersal Risk of Invasive Alien Plant Species," In *Recent Advances in Applications of Computational and Fuzzy Mathematics.* (Springer 2018) 109-145 with permission.)

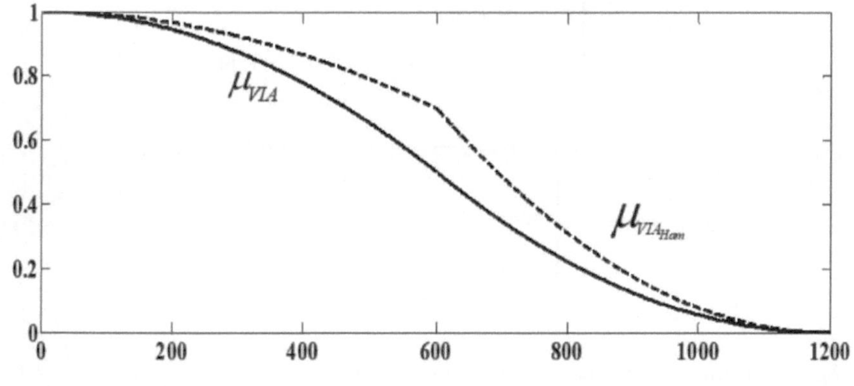

Figure 8.2 Original and modified membership functions of *VIA* in Model I-Hamacher. (Modified from H.O.W. Peiris, S. Chakraverty, S.S.N. Perera, S.S.N and S.M.W. Ranwala, "Modelling Dispersal Risk of Invasive Alien Plant Species," In *Recent Advances in Applications of Computational and Fuzzy Mathematics.* (Springer 2018) 109-145 with permission)

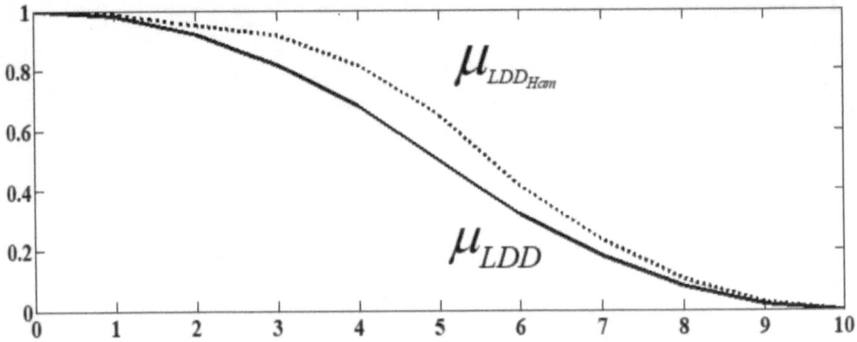

Figure 8.3 Original and modified membership functions of *LDD* in Model I-Hamacher. (Modified from H.O.W. Peiris, S. Chakraverty, S.S.N. Perera, S.S.N and S.M.W. Ranwala, "Modelling Dispersal Risk of Invasive Alien Plant Species", In *Recent Advances in Applications of Computational and Fuzzy Mathematics.* (Springer 2018) 109-145 with permission.)

$$\mu_{VIA_{Ham}}(x) = \begin{cases} 1 & \text{for } x < 3 \\ 1 - 2\left[\dfrac{(x-3)^2}{2376060}\right] & \text{for } 3 \le x < 602 \\ 2\left[\dfrac{(1200-x)^2}{1.0247e+006}\right] & \text{for } 602 \le x \le 1200 \\ 0 & \text{for } x > 1200 \end{cases} \tag{8.10}$$

$$\mu_{LDD_{Ham}}(x) = \begin{cases} 1 - 2\left[\dfrac{x^2}{160}\right] & \text{for } 0 \le x < 2 \\ 0.95 - 2\left[\dfrac{(x-2)^2}{60}\right] & \text{for } 2 \le x < 5 \\ 2\left[\dfrac{(10-x)^2}{77}\right] & \text{for } 5 \le x \le 10 \end{cases} \tag{8.11}$$

The algorithm for the model is presented below:

- vector p denotes the original performance values of the four parameters NSF, ASR, VIA, and LDD of a particular plant species.
- vector q denotes the membership values corresponding to Eqs. (8.5) to (8.8) of the four parameters NSF, ASR, VIA, and LDD of a particular plant species.
- vector r denotes the membership values corresponding to Eqs. (8.9) to (8.11) of the three parameters ASR, VIA, and LDD of a particular plant species.

- $Hamacher_{p'}$ denotes the Hamacher operator with parameter p' as in Eq.(3.13).

Enter vector $p = [x_{NSF}, x_{ASR}, x_{VIA}, x_{LDD}]$
Enter vector $q = [\mu_{NSF}(x), \mu_{ASR}(x), \mu_{VIA}(x), \mu_{LDD}(x)]$
Enter vector $r = [\mu_{ASR_{Ham}}(x), \mu_{VIA_{Ham}}(x), \mu_{LDD_{Ham}}(x)]$
If $x_{NSF} \leq 200$ (**Category I**)
If $ASR \leq 20000$ and $VIA \leq 180$ (**Category II**)
$Hamacher_{0.9}\left(\mu^{6}_{NSF}(x), \mu_{ASR_{Ham}}(x), \mu^{73.5}_{VIA_{Ham}}(x), \mu_{LDD_{Ham}}(x)\right)$
Else
$Hamacher_{0.9}\left(\mu^{6}_{NSF}(x), \mu_{ASR_{Ham}}(x), \mu_{VIA_{Ham}}(x), \mu_{LDD_{Ham}}(x)\right)$
End
Else if $ASR \leq 100000$ and $NSF \leq 100$ (**Category III**)
$Hamacher_{0.9}\left({\mu_{NSF}}^{0.5}(x), \mu_{ASR_{Ham}}(x), \mu^{73.5}_{VIA_{Ham}}(x), \mu_{LDD_{Ham}}(x)\right)$
Else if $ASR \geq 10000$ and $LDD \geq 100$ (**Category IV**)
$Hamacher_{0.9}\left(\mu_{NSF}(x), \mu_{ASR}(x), \mu_{VIA}(x), \mu^{0.1}_{LDD_H}(x)\right)$
Else (**Category V**)
$Hamacher_{0.9}\left(\mu_{NSF}(x), \mu_{ASR_{Ham}}(x), \mu_{VIA_{Ham}}(x), \mu_{LDD_{Ham}}(x)\right)$
End
End
End

It is clear that the above alogirithm is executed over five plant categories with different parametric values of p'. Also it can be seen that importance weights of parameters have also been changed over the plant category. Below we briefly explain how the dispersal risk evaluation is processed in each plant category.

Category I exposes the species with *NSF* less than or equal 200. Here the membership value of μ_{NSF} is downgraded by 6 (see Figure 8.4). For the remaining parameters the modified membership functions from Eqs. 8.9 to 8.11 are used.

Category II exposes the species with $ASR \leq 20000$ and $VIA \leq 180$ while satisfying the conditions stated in Category I. In that case, the membership value of $\mu_{VIA_{Ham}}$ (see Eq. 8.10) is downgraded by 73.5 while using the modified membership functions for the remaining parameters as in Eqs. (8.9) and (8.11).

Category III: The species which produces *ASR* greater than or equal to 100,000 with *SF* less than or equal to 100 belongs to this category. Here the upgraded membership value of μ_{NSF} is 0.5 with modified membership functions from Eqs. (8.9) to (8.11) for the remaining parameters.

Category IV: The species with $ASR < 10000$ and $LDD \geq 6$ belongs in this category. The upgraded value of $\mu_{LDD_{Ham}}$ is 0.1, aggregated with the original fuzzy sets as in Eqs (8.5) to (8.7).

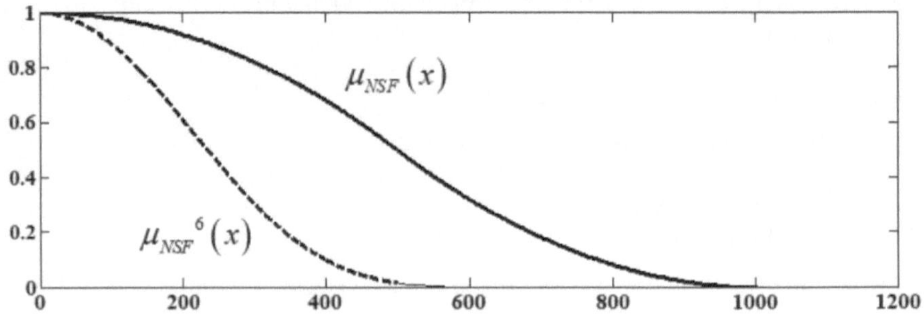

Figure 8.4 Original and modified membership functions of *NSF* in Model I-Hamacher.

Category V: This is the last category which is composed of the species which does not belong to any of the above categories. In that case, the original membership function of μ_{SF} is combined with the modified membership functions of the remaining parameters.

The test results are presented in Table 8.1. Here we have validated the model using known invasive and non-invasive species. The chosen invasive species show invasiveness mainly through other factors than the dispersal factors. The reason for this selection is to check whether the species which do not pose considerable dispersal risk should present low risk level. The results are presented in Table 8.2.

8.4 MODEL II – YAGER *t*-NORM OPERATOR

In this model, the parameterized Yager *t*-norm operator as defined in Eq.(3.15) has been used as the aggregation operator. Following Steps 2 to 4 as in Section 8.2, the invasive plants are classified into three categories. All the membership functions of parameters used in each category remain as original functions as defined in Eqs (8.5), (8.7), and (8.8) except for the parameter *ASR*. In Categories I and II, the membership function of *ASR* has been modified as given in Eq.(8.12) (see Figure 8.5). The algorithm of the model is presented as follows:

- vector p denotes the original numerical values of the four parameters NSF, ASR, VIA, and LDD of a particular plant species.
- vector q denotes the membership values corresponding to Eqs. (8.5) to (8.8) of the four parameters NSF, ASR, VIA, and LDD of a particular plant species.
- vector r denotes the membership values corresponding to Eq.(8.12) of the parameter ASR of a particular plant species.
- $Yager_{p'}$ denotes the Yager operator with parameter p' as in Eq. (3.15).

Table 8.1

Test results - Model I Adapted from H.O.W. Peiris, S. Chakraverty, S.S.N. Perera, S.S.N and S.M.W. Ranwala, "Modelling Dispersal Risk of Invasive Alien Plant Species", In Recent *Advances in Applications of Computational and Fuzzy Mathematics*. (Springer 2018) 109–145.

Invasive Species	Dispersal Risk Level (Model I - Hamacher)	Dispersal Risk Level (NRA)
Alternanthera philoxeroides	*Low*	*Low*
Clidemia hirta	*High*	*High*
Miconia calvescens	*Extremely High*	*Extremely High*
Alstonia macrophylla	*Very High*	*Very High*
Annona glabra	*Very Low*	*Very Low*
Clusia rosea	*Medium*	*Medium*
Dillenia suffructicosa	*Very Low*	*Very Low*
Ageratina riparia	*High*	*High*
Mimosa invisa	*High*	*High*
Myroxylon balsamum	*Medium*	*Medium*
Tithonia diversiflora	*Low*	*Low*
Mikania micrantha	*High*	*High*
Prosopis juliflora	*High*	*High*
Ulex europaeus	*Medium*	*Medium*
Mimosa pigra	*High*	*High*
Chromolaena odorata	*High*	*High*
Parthenium hysterophorus	*High*	*High*
Lantana camara	*Medium*	*Medium*
Imperata cylindrical	*High*	*High*

Table 8.2

Validation results - Model I Adapted from H.O.W. Peiris, S. Chakraverty, S.S.N. Perera, S.S.N and S.M.W. Ranwala, "Modelling Dispersal Risk of Invasive Alien Plant Species", In *Recent Advances in Applications of Computational and Fuzzy Mathematics*. (Springer 2018) 109-145.

Category of Species	Species	Dispersal Risk Level (Model I - Hamacher)	Dispersal Risk Level (NRA)
Invasive	*Sphagneticola trilobata*	*Low*	*Medium*
	Cuscuta campestris	*Low*	*Medium*
	Pueraria Montana	*Low*	*Low*
Non-Invasive	*Cassia fistula*	*Low*	*Low*
	Cissus rotundifolia	*Very Low*	*Low*
	Hedychium gardnerianum	*Very Low*	*Very Low*
	Magnefera indica	*Very Low*	*Low*

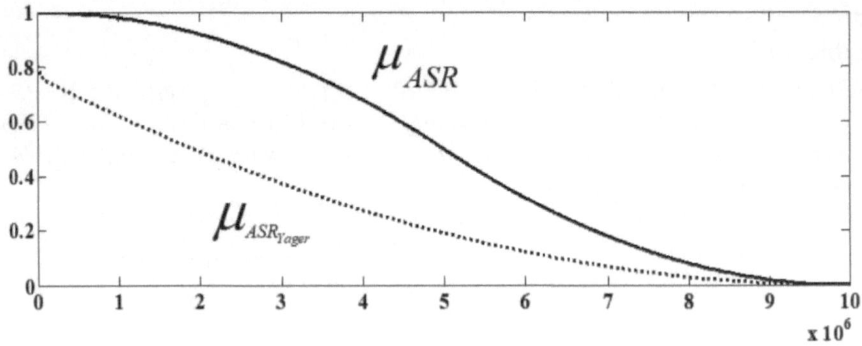

Figure 8.5 Original and modified membership functions of *ASR* in Model II-Yager.

Enter vector $p = [x_{NSF}, x_{ASR}, x_{VIA}, x_{LDD}]$
Enter vector $q = [\mu_{NSF}(x), \mu_{ASR}(x), \mu_{VIA}(x), \mu_{LDD}(x)]$
Enter vector $r = [\mu_{ASR_{Yager}}(x)]$
If $x_{NSF} \leq 200$ (**Category I**)
If $ASR \geq 10000$ and $1 \leq VIA \leq 300$
$Yager_{p'=1.1}\left(\mu_{NSF}^{10.9}(x), \mu_{ASR_{Yager}}(x), \mu_{VIA}(x), \mu_{LDD}(x)\right)$
Else
$Yager_{p'=1.1}\left(\mu_{NSF}(x), \mu_{ASR_{Ham}}(x), \mu_{VIA_{Ham}}(x), \mu_{LDD_{Ham}}(x)\right)$
End
Else if $ASR \geq 10000$ and $LDD \geq 100$ (**Category II**)
$Yager_{p'=1.1}\left(\mu_{NSF}(x), \mu_{ASR_{Yager}}(x), \mu_{VIA}(x), \mu_{LDD}^{0.1}(x)\right)$
Else (**Category III**)
$Yager_{p'=2.1}\left(\mu_{NSF}(x), \mu_{ASR}(x), \mu_{VIA}(x), \mu_{LDD}(x)\right)$
End
End
End

According to the algorithm, it executes with three plant categories and different values for p' as inputs. It can be seen that for some parameters in Categories I to II, the important weights are incorporated.

Following we explain the evaluation process of dispersal risk in each category in detail.

Category I: The category consists of two steps. The first step executes with the plants which pose $ASR \geq 10000$ with $1 \leq VIA \leq 300$. Here the modified $\mu_{ASR_{Yager}}$ as defined in Eq.(8.12) with original membership functions for the remaining parameters are used for the aggregation. When it turns to the next step, plants that belongs to the first category should preserve $NSF \geq 200$ additionally. In that step the membership function μ_{NSF} is downgraded by 10.9. The suitable parameter value $\acute{p}$ for this aggregation has been identified as 1.1 after simulating with several values greater than one.

Category II: This category contains plant species that do not belong to Category I with $LDD \geq 5$. The modified membership function of $\mu_{ASR_{Yager}}$ as defined in Eq. (8.12) with an upgraded membership function of μ_{LDD} by 0.1 are considered here with the original fuzzy sets for the remaining parameters. Here the $\acute{p}$ value is 1.1.

Category III: This category proceeds with species which do not belong to either of the above categories. Here only the original fuzzy set for the each parameter is used for the aggregation by setting the $\acute{p}$ value as 2.1.

Model II has been validated using the same set of species as was used in Model I. The test results and validation results are illustrated in Tables 8.3 and 8.4.

$$\mu_{ASR_{Yager}}(x) = \begin{cases} 1-2\left[\dfrac{(x-0)^2}{1.0526e+009}\right]+0.75 & \text{for } 0 \leq x < 10000 \\ 2\left[\dfrac{(100000-x)^2}{2.7e+011}\right]+0.75 & \text{for } 10000 \leq x < 100000 \\ 2\left[\dfrac{(10^7-x)^2}{2.613e+014}\right] & \text{for } 100000 \leq x \leq 10^7. \end{cases} \quad (8.12)$$

8.5 MODEL III – DOMBI *t*-NORM OPERATOR

Here the parameterized Dombi t–norm operator has been chosen as the aggregation operator. After following up the procedure stated in Section 8.2, it is found that the model execution runs through five plant categories. Except in Category V, other categories execute with original fuzzy sets which have been defined initially. The alogirthm for this model is given as below:

- vector p denotes the original numerical values of the four parameters NSF, ASR, VIA, and LDD of a particular plant species.
- vector q denotes the membership values corresponding to Eqs. (8.5) to (8.8) of the four parameters NSF, ASR, VIA, and LDD of a particular plant species.
- vector r denotes the membership values corresponding to Eq.(8.12) of the parameter ASR of a particular plant species.
- $Dombi_{p'}$ denotes the Dombi operator with parameter p' as in Eq. (3.19).

Table 8.3

Test results - Model II Adapted from H.O.W. Peiris, S. Chakraverty, S.S.N. Perera, S.S.N and S.M.W. Ranwala, "Modelling Dispersal Risk of Invasive Alien Plant Species", In *Recent Advances in Applications of Computational and Fuzzy Mathematics*. (Springer 2018) 109-145.

Invasive Species	Dispersal Risk Level (Model II - Yager)	Dispersal Risk Level (NRA)
Alternanthera philoxeroides	*Low*	*Low*
Clidemia hirta	*High*	*High*
Miconia calvescens	*Extremely High*	*Extremely High*
Alstonia macrophylla	*Very High*	*Very High*
Annona glabra	*Very Low*	*Very Low*
Clusia rosea	*Medium*	*Medium*
Dillenia suffructicosa	*Very Low*	*Very Low*
Ageratina riparia	*Low*	*High*
Mimosa invisa	*High*	*High*
Myroxylon balsamum	*Medium*	*Medium*
Tithonia diversiflora	*Low*	*Low*
Mikania micrantha	*Low*	*High*
Prosopis juliflora	*High*	*High*
Ulex europaeus	*Medium*	*Medium*
Mimosa pigra	*Very High*	*High*
Chromolaena odorata	*Medium*	*High*
Parthenium hysterophorus	*Medium*	*High*
Lantana camara	*Medium*	*Medium*
Imperata cylindrical	*Medium*	*High*

Table 8.4

Validation results - Model II Adapted from H.O.W. Peiris, S. Chakraverty, S.S.N. Perera, S.S.N and S.M.W. Ranwala, "Modelling Dispersal Risk of Invasive Alien Plant Species", In *Recent Advances in Applications of Computational and Fuzzy Mathematics*. (Springer 2018) 109-145.

Category of Species	Species	Dispersal Risk Level (Model II - Yager)	Dispersal Risk Level (NRA)
Invasive	*Sphagneticola trilobata*	*Low*	*Medium*
	Cuscuta campestris	*Low*	*Medium*
	Pueraria Montana	*Low*	*Low*
Non-Invasive	*Cassia fistula*	*Low*	*Low*
	Cissus rotundifolia	*Low*	*Low*
	Hedychium gardnerianum	*Very Low*	*Very Low*
	Magnefera indica	*Very Low*	*Low*

Enter vector $p = [x_{NSF}, x_{ASR}, x_{VIA}, x_{LDD}]$
Enter vector $q = [\mu_{NSF}(x), \mu_{ASR}(x), \mu_{VIA}(x), \mu_{LDD}(x)]$
Enter vector $r = [\mu_{ASR_{Yager}}(x)]$
If $x_{NSF} \leq 200$ (**Category I**)
If $LDD \geq 4$ and $VIA \leq 300$
$Dombi_{p'=0.29}(\mu_{NSF}(x), \mu_{ASR}(x), \mu_{VIA}(x), \mu_{LDD}(x))$
Else if $LDD < 4$ and $NSF \geq 200$ and $ASR < 50000$ (**Category II**)
$Dombi_{p'=0.23}(\mu_{NSF}(x), \mu_{ASR}(x), \mu_{VIA}(x), \mu_{LDD}(x))$
Else if $LDD < 4$ and $NSF \geq 200$ and $ASR > 100000$ (**Category III**)
$Dombi_{p'=0.39}(\mu_{NSF}(x), \mu_{ASR}(x), \mu_{VIA}(x), \mu_{LDD}(x))$
Else if $ASR < 1000$ and $LDD \geq 6$ (**Category IV**)
$Dombi_{p'=0.5}\left(\mu_{NSF}(x), \mu_{ASR}(x), \mu_{VIA}(x), \mu_{LDD_{0.1}}(x)\right)$
Else if $VIA \geq 600$ (**Category V**)
$Dombi_{p'=1.2}\left(\mu_{NSF}(x), \mu_{ASR_{Yager}}(x), \mu_{VIA}(x), \mu_{LDD}(x)\right)$
Else
$Dombi_{p'=0.5}\left(\mu_{NSF}(x), \mu_{ASR_{Yager}}(x), \mu_{VIA}(x), \mu_{LDD}(x)\right)$
End
End
End
End
End

Below we discuss explicitly how each category mentioned in the algorithm executes:

Category I: This category runs through the species with $LDD \geq 4$ and $VIA \leq 300$. Here aggregation of parameters is followed by original fuzzy sets with $\acute{p} = 0.23$.

Category II: This part executes with the species with $LDD < 4$ and $NSF \geq 200$ and $ASR < 50000$. Here the only difference compared to Category I is the p' value which is 0.23.

Category III: The species with $LDD < 4$ and $NSF \geq 200$ and $ASR > 100000$ belongs to this category. This part executes the same way as the first category with $p' = 0.39$.

Category IV: For Category IV, the species should preserve $ASR < 1000$ and $LDD \geq 6$. Here the original membership function of μ_{LDD} is upgraded by 0.1 and aggregates with the original fuzzy sets of remaining parameters with $p' = 0.5$.

Category V: This is the final category which executes with the species that does not belong to either above category. Here $\mu_{ASR_{Yager}}$ is defined in Eq. (8.12) and aggregated with original fuzzy sets of remaining parameters with $p' = 1.2$ if the species is with $VIA \geq 600$; otherwise use $p = 0.5$.

Table 8.5

Test results - Model III Adapted from H.O.W. Peiris, S. Chakraverty, S.S.N. Perera, S.S.N and S.M.W. Ranwala, "Modelling Dispersal Risk of Invasive Alien Plant Species", In *Recent Advances in Applications of Computational and Fuzzy Mathematics*. (Springer 2018) 109-145.

Invasive Species	Dispersal Risk Level (Model III - Dombi)	Dispersal Risk Level (NRA)
Alternanthera philoxeroides	*Low*	*Low*
Clidemia hirta	*High*	*High*
Miconia calvescens	*Extremely High*	*Extremely High*
Alstonia macrophylla	*Very High*	*Very High*
Annona glabra	*Low*	*Very Low*
Clusia rosea	*Medium*	*Medium*
Dillenia suffructicosa	*Very Low*	*Very Low*
Ageratina riparia	*Medium*	*High*
Mimosa invisa	*High*	*High*
Myroxylon balsamum	*Low*	*Medium*
Tithonia diversiflora	*Low*	*Low*
Mikania micrantha	*Medium*	*High*
Prosopis juliflora	*Very High*	*High*
Ulex europaeus	*Medium*	*Medium*
Mimosa pigra	*Very High*	*High*
Chromolaena odorata	*Medium*	*High*
Parthenium hysterophorus	*Low*	*High*
Lantana camara	*Medium*	*Medium*
Imperata cylindrical	*High*	*High*

The model has been validated as in the previous model and the test results and validation results are shown in Tables 8.5 and 8.6.

8.6 MODEL IV—DUBOIS AND PRADE t—NORM OPERATOR

This model operates with the parameterized Dubois and Prade t - norm fuzzy operator. Unlike the previous models, this model has shown much flexibility with the original fuzzy sets which are defined in Eqs (8.5) to (8.8). The model runs through five plant categories and the algorithm is presented below.

- vector p denotes the original numerical values of the four parameters NSF, ASR, VIA, and LDD of a particular plant species.
- vector q denotes the membership values corresponding to Eqs. (8.5) to (8.8) of the four parameters NSF, ASR, VIA, and LDD of a particular plant species.
- $Dubois_{p'}$ denotes the Dubois and Prade operator with parameter p as in Eq. (3.17).

Table 8.6

Validation results - Model III Adapted from H.O.W. Peiris, S. Chakraverty, S.S.N. Perera, S.S.N and S.M.W. Ranwala, "Modelling Dispersal Risk of Invasive Alien Plant Species", In *Recent Advances in Applications of Computational and Fuzzy Mathematics*. (Springer 2018) 109-145.

Category of Species	Species	Dispersal Risk Level (Model III - Dombi)	Dispersal Risk Level (NRA)
Invasive	*Sphagneticola trilobata*	*Very Low*	*Medium*
	Cuscuta campestris	*Very Low*	*Medium*
	Pueraria Montana	*Very Low*	*Low*
Non-Invasive	*Cassia fistula*	*Very Low*	*Low*
	Cissus rotundifolia	*Very Low*	*Low*
	Hedychium gardnerianum	*Unlikely*	*Very Low*
	Magnefera indica	*Very Low*	*Low*

Enter vector $p = [x_{NSF}, x_{ASR}, x_{VIA}, x_{LDD}]$
Enter vector $q = [\mu_{NSF}(x), \mu_{ASR}(x), \mu_{VIA}(x), \mu_{LDD}(x)]$
If $ASR \geq 100000$ **(Category I)**
$Dubois_{p'=1.1}(\mu_{NSF}(x), \mu_{ASR}(x), \mu_{VIA}(x), \mu_{LDD}(x))$
Else if $ASR < 50000$ **(Category II)**
$Dubois_{p'=1}(\mu_{NSF}(x), \mu_{ASR}(x), \mu_{VIA}(x), \mu_{LDD}(x))$
Else if $ASR < 50000$ and $VIA \geq 600$ **(Category III)**
$Dubois_{p'=0.9}(\mu_{NSF}(x), \mu_{ASR}(x), \mu_{VIA}(x), \mu_{LDD}(x))$
Else if $ASR < 50000$ and $NSF \geq 200$ **(Category IV)**
$Dubois_{p'=2}(\mu_{NSF}(x), \mu_{ASR}(x), \mu_{VIA}(x), \mu_{LDD_{0.1}}(x))$
Else if $ASR < 10000$ and $LDD \geq 6$ **(Category V)**
$Dubois_{p'=0.1}(\mu_{NSF}(x), \mu_{ASR_{Yager}}(x), \mu_{VIA}(x), \mu_{LDD_{0.1}}(x))$
End
End
End
End
End

The execution of each category mentioned in the above alogirithm is explained below.

Category I: The species belongs to the first category that preserves $ASR \geq 100000$ and runs with original fuzzy sets of the four parameters with $p' = 1.1$.

Category II: This group executes species with $ASR < 50000$ by aggregating original fuzzy sets with $p' = 1$.

Category III: The species with $ASR < 50000$ and $VIA \geq 600$ belongs to this category which executes the same as the second category with $p' = 0.9$.

Table 8.7

Test results - Model IV Adapted from H.O.W. Peiris, S. Chakraverty, S.S.N. Perera, S.S.N and S.M.W. Ranwala, "Modelling Dispersal Risk of Invasive Alien Plant Species", In *Recent Advances in Applications of Computational and Fuzzy Mathematics*. (Springer 2018) 109-145.

Invasive Species	Dispersal Risk Level (Model IV - Dubois)	Dispersal Risk Level (NRA)
Alternanthera philoxeroides	*Low*	*Low*
Clidemia hirta	*High*	*High*
Miconia calvescens	*Extremely High*	*Extremely High*
Alstonia macrophylla	*Very High*	*Very High*
Annona glabra	*Very Low*	*Very Low*
Clusia rosea	*Medium*	*Medium*
Dillenia suffructicosa	*Very Low*	*Very Low*
Ageratina riparia	*Medium*	*High*
Mimosa invisa	*High*	*High*
Myroxylon balsamum	*Medium*	*Medium*
Tithonia diversiflora	*Low*	*Low*
Mikania micrantha	*Medium*	*High*
Prosopis juliflora	*High*	*High*
Ulex europaeus	*High*	*Medium*
Mimosa pigra	*High*	*High*
Chromolaena odorata	*Medium*	*High*
Parthenium hysterophorus	*High*	*High*
Lantana camara	*Medium*	*Medium*
Imperata cylindrical	*High*	*High*

Category IV: This group is related with the species where $ASR < 50000$ and $NSF \geq 200$ and executes the same way as previous categories with $p' = 2$.

Category V: This category is for the species with $ASR < 10000$ and $LDD \geq 6$. Here the original membership function μ_{LDD} is upgraded by 0.1 and the p' value is 0.1.

The same validation process as was used in previous models has been conducted with this model. The test results and validation results are displayed in Tables 8.7 and 8.8.

8.7 DISCUSSION MODEL I—MODEL IV

According to the algorithms of the four models mentioned above, each model executes with several categories divided on the basis of the performances of plant species viz. four dispersal related parameters/factors.
Specifically, Model II which has been developed based on Yager that enables the least number of categories compared to other models.

If we compare the performances of Models I and II, it can be seen that the risk level of each species in Model I is compatible with NRA level. In contrast, except

Table 8.8

Validation results - Model IV Adapted from H.O.W. Peiris, S. Chakraverty, S.S.N. Perera, S.S.N and S.M.W. Ranwala, "Modelling Dispersal Risk of Invasive Alien Plant Species", In *Recent Advances in Applications of Computational and Fuzzy Mathematics.* (Springer 2018) 109-145.

Category of Species	Species	Dispersal Risk Level (Model IV - Dubois)	Dispersal Risk Level (NRA)
Invasive	*Sphagneticola trilobata*	*Low*	*Medium*
	Cuscuta campestris	*Low*	*Medium*
	Pueraria Montana	*Low*	*Low*
Non-Invasive	*Cassia fistula*	*Low*	*Low*
	Cissus rotundifolia	*Low*	*Low*
	Hedychium gardnerianum	*Very Low*	*Very Low*
	Magnefera indica	*Very Low*	*Low*

species *Ageratina riparia, Mikania micrantha, Mimosa pigra, Chromolaena odorata, Parthenium hysterophorus, Imperata cylindrica*, the risk levels of the remaining species are compatible with NRA values in Model II.

As we mentioned earlier, the invasive species chosen for the validation show low performances on dispersal and higher performances on other related risk factors. For example, species *Sphagneticola trilobata* and *Cuscuta campestris* take *Medium* risk from NRA, although one may expect a much lower level risk for this category. According to Tables 8.2 and 8.4, it can be clearly seen that risk levels of these invasive species show lower risk levels than the NRA level as expected.

Besides, species in the non-invasive category show low risk levels compared to the NRA level. Therefore, considering the test results and validation, one may conclude that Model I developed based on the Hamacher operator better enables capturing the dispersal risk than Model II.

Now let us compare the test results of Model III and Model IV. According to Table 8.5, one may see that species risk levels obtained from Model III are compatible with the NRA level except for *Annona glabra, Ageratina riparia, Myroxylon balsamum, Mikania micrantha, Prosopis juliflora, Mimosa pigra, Chromolaena odorata*, and *Parthenium hysterophorus*. As presented in Table 8.7, risk levels of species evaluated through Model IV are compatible with NRA except for *Ageratina riparia*.

Further, one may note that validation results of Models III and IV as in Tables 8.6 and 8.8 are quite similar to the validation results of Models I and II.

These four models execute in different ways as some need original fuzzy sets whereas others need modified fuzzy sets. Comparing the test and validation results of each model over the other, it can be concluded that Model I, which has been

Table 8.9

Importance weights for dispersal risk parameters Adapted from H.O.W. Peiris, S. Chakraverty, S.S.N. Perera, S.S.N and S.M.W. Ranwala, "Modelling Dispersal Risk of Invasive Alien Plant Species", In *Recent Advances in Applications of Computational and Fuzzy Mathematics*. (Springer 2018) 109-145.

Parameter	Weights Chang's Method	Weights Buckly's Method
Number of seeds per fruit	0.2884	0.23924
Annual seed production per m^2	0.2785	0.26698
Viability of seeds	0.2504	0.29315
Long distance dispersal strength	0.1827	0.20062

developed based on the Hamacher operator gives more significant results than the other models.

8.8 EVALUATING IMPORTANCE WEIGHTS USING FUZZY ANALYTICAL HIERARCHY TECHNIQUES

Here we focus on evaluating the importance weights of the four dispersal risk parameters by following Chang's and Buckly's methods described in Section 4.5.2.

First, a questionnaire has been prepared to obtain the all possible pairwise comparisons among four parameters; NSF, ASR, VIA, and LDD. A group of three plant science experts have filled out the questionnaire individually.

Using Chang's method and Bucky's method as given in Eqs. (4.11) to (4.18), two sets of importance weights for the parameters NSF, ASR, VIA, and LDD have been obtained. These weights are presented in Table 8.9.

In the next subsection, the models based on the importance weights are explained.

8.8.1 MODELS – PROPOSED METHOD I

Here we develop models to evaluate the dispersal risk of IAPS using Method I mentioned in Eq. (4.19). Accordingly, P_1, P_2, P_3, and P_4 in Eq. (4.19) denote the parameters NSF, ASR, VIA, LDD respectively and w_i denotes the importance weights of these parameters. Therefore, we have separately applied the weights obtained from Chang's and Buckly's methods. As described in Figure 4.8., initial p_j values are assigned. This process has been executed back and forth until the output is considerably close to the NRA levels. It has been identified that the model is executed with two plant categories as follows:

Table 8.10

Test results - Model I Adapted from H.O.W. Peiris, S. Chakraverty, S.S.N. Perera, S.S.N and S.M.W. Ranwala, "Novel Fuzzy Based Model on Analysis of Invasiveness due to Dispersal related Traits of Plants," *Annals of Fuzzy Mathematics and Informatics*, 13, no(3) (2017): 6-14.

Invasive Species	Model I - Chang's Method Risk Level	NRA Risk Level
Alternanthera philoxeroides	*Very Low*	*Low*
Clidemia hirta	*High*	*High*
Miconia calvescens	*High*	*Extremely High*
Alstonia macrophylla	*Medium*	*Very High*
Annona glabra	*Medium*	*Very Low*
Clusia rosea	*Medium*	*Medium*
Dillenia suffructicosa	*Low*	*Very Low*
Ageratina riparia	*Low*	*High*
Mimosa invisa	*High*	*High*
Myroxylon balsamum	*Medium*	*Medium*
Tithonia diversiflora	*Low*	*Low*
Mikania micrantha	*Low*	*High*
Prosopis juliflora	*High*	*High*
Ulex europaeus	*High*	*Medium*
Mimosa pigra	*High*	*High*
Chromolaena odorata	*High*	*High*
Parthenium hysterophorus	*Low*	*High*
Lantana camara	*High*	*Medium*
Imperata cylindrica	*Very High*	*High*

Category I: A plant whose $ASR \leq 20000$ and $VIA \leq 3$yrs
Here the model is given by Eq. (8.13):

$$Output(x) = w_1\mu_{NSF}^{\frac{1}{2}} + w_2\mu_{ASR}^{\frac{1}{2}} + w_3\mu_{VIA}^{\frac{1}{12.5}} + w_4\mu_{LDD}^{\frac{1}{3}}, x\varepsilon X. \tag{8.13}$$

Category II: A plant does not belong to category I
In this case the model may be given as Eq. (8.14):

$$Output(x) = w_1\mu_{NSF}^{\frac{1}{2}} + w_2\mu_{ASR}^{\frac{1}{10}} + w_3\mu_{VIA}^{\frac{1}{120}} + w_4\mu_{LDD}^{\frac{1}{3}}, x\varepsilon X. \tag{8.14}$$

8.8.2 DISCUSSION – MODELS I AND II

According to Table 8.9, the highest importance weight is for the parameter: number of seeds per fruit (NSF) and the lowest weight is for the long distance dispersal strength parameters (LDD) in Chang's method. Besides, Buckly's method gives the highest weight for the viability of seeds (VIA) factor whereas the lower weight is for the long distance dispersal (LDD).

Table 8.11

Test results - Model II Adapted from H.O.W. Peiris, S. Chakraverty, S.S.N. Perera, S.S.N and S.M.W. Ranwala, "Novel fuzzy based model on analysis of invasiveness due to dispersal related traits of plants," *Annals of Fuzzy Mathematics and Informatics*, 13, no(3) (2017): 6-14.

Invasive Species	Model II - Buckly's Method Risk Level	NRA Risk Level
Alternanthera philoxeroides	*Low*	*Low*
Clidemia hirta	*High*	*High*
Miconia calvescens	*High*	*Extremely High*
Alstonia macrophylla	*Medium*	*Very High*
Annona glabra	*Medium*	*Very Low*
Clusia rosea	*Medium*	*Medium*
Dillenia suffructicosa	*Low*	*Very Low*
Ageratina riparia	*Low*	*High*
Mimosa invisa	*High*	*High*
Myroxylon balsamum	*Medium*	*Medium*
Tithonia diversiflora	*Low*	*Low*
Mikania micrantha	*Low*	*High*
Prosopis juliflora	*High*	*High*
Ulex europaeus	*High*	*Medium*
Mimosa pigra	*High*	*High*
Chromolaena odorata	*High*	*High*
Parthenium hysterophorus	*Low*	*High*
Lantana camara	*High*	*Medium*
Imperata cylindrica	*Very High*	*High*

Table 8.12

Validation results - Model I Adapted from H.O.W. Peiris, S. Chakraverty, S.S.N. Perera, S.S.N and S.M.W. Ranwala, "Novel fuzzy based model on analysis of invasiveness due to dispersal related traits of plants," *Annals of Fuzzy Mathematics and Informatics*, 13, no(3) (2017): 6-14.

Category of Species	Species	Model I-Chang's Risk Level	NRA Risk Level
Invasive	*Sphagneticola trilobata*	*Low*	*Medium*
	Cuscuta campestris	*Low*	*Medium*
	Pueraria montana	*Low*	*Low*
Non-Invasive	*Cassia fistula*	*Low*	*Low*
	Cissus rotundifolia	*Low*	*Low*
	Hedychium gardnerianum	*Very Low*	*Very Low*
	Mangifera indica	*Very Low*	*Low*

Table 8.13

Validation results – Model II Adapted from H.O.W. Peiris, S. Chakraverty, S.S.N. Perera, S.S.N and S.M.W. Ranwala, "Novel fuzzy based model on analysis of invasiveness due to dispersal related traits of plants," *Annals of Fuzzy Mathematics and Informatics,* 13, no(3) (2017): 6-14.

Category of Species	Species	Model II-Buckly's Risk Level	NRA Risk Level
Invasive	*Sphagneticola trilobata*	*Medium*	*Medium*
	Cuscuta campestris	*Medium*	*Medium*
	Pueraria montana	*Low*	*Low*
Non-Invasive	*Cassia fistula*	*Low*	*Low*
	Cissus rotundifolia	*Low*	*Low*
	Hedychium gardnerianum	*Very Low*	*Very Low*
	Mangifera indica	*Very Low*	*Low*

Let us first compare the test results of Models I and II as illustarted in Tables 8.10 to 8.11. In Model I, risk values of species can be seen such that *Clidemia hirta, Clusia rosea, Mimosa invisa, Myroxylon balsamum, Tithonia diversiflora, Prosopis juliflora, Mimosa pigra, Chromolaena odorata, Miconia calvescens* are compatible with the NRA values. On the other hand, with species like *Alternanthera philoxeroides, Alstonia macrophylla, Ageratina riparia, Mikania micrantha, Parthenium hysterophorus,* their risk values stand behind the NRA values. Also it can be clearly seen that test results of Model II behave the same as Model I.

If we analyze the validation results of Model II as presented in Table 8.12, the NRA risk levels of invasive species *Sphagneticola trilobata* and *Cuscuta campestris,* that is "*Medium*" are compatible with the model output. These risk levels are considerably high, as we expect a low risk in reality.
However, in the validation results of Model I in Table 8.13, the same species mentioned above take the risk level of "*Low*" which is one level behind the NRA level as expected. The risk levels of non-invasive species except *Magnefera indica* in both models are compatible with the NRA levels which are considerably low risk levels. The species *Magnefera indica* shows "*Very Low*" risk level compared to the "*Low*" NRA level in both models.

After analyzing the test and validation results, one may conclude that the ability to capture invasiveness is greater in Model I than in Model II. In the next section, we discuss the models which have been developed by combining Chang's and Buckly's methods.

8.9 MODELS – PROPOSED METHOD II

In this section, we develop models to evaluate the dispersal risk of IAPS using the proposed Method II in Subsection 4.5.4. As in the proposed Method II, the new

Table 8.14
Grade of importance weights W_T for Dispersal related factors Adapted from H.O.W. Peiris, S. Chakraverty, S.S.N. Perera, S.S.N and S.M.W. Ranwala, "Modelling Dispersal Risk of Invasive Alien Plant Species", In *Recent Advances in Applications of Computational and Fuzzy Mathematics.* (Springer 2018) 109-145.

λ value	W_T for dispersal factors			
	SF	*ASR*	*VIA*	*LDD*
0.1	0.244156	0.268132	0.288875	0.198828
0.2	0.249072	0.269284	0.2846	0.197036
0.3	0.253988	0.270436	0.280325	0.195244
0.4	0.258904	0.271588	0.27605	0.193452
0.5	0.26382	0.27274	0.271775	0.19166
0.6	0.268736	0.273892	0.2675	0.189868
0.7	0.273652	0.275044	0.263225	0.188076
0.8	0.278568	0.276196	0.25895	0.186284
0.9	0.283484	0.277348	0.254675	0.184492

weight W_T need to be generated using Chang's and Buckly's methods. Following Eq. (4.20), W_T weights have been obtained by generating λ values between 0 and 1. The results are presented in Table 8.14.

Here we develop two models based on the weights W_T which are presented in the following sections.

8.9.1 CONSTRUCTION OF MODELS III AND IV

Here Models III and IV have been constructed using different sets of membership values for the μ_{P_n}'s in Eq. (4.21). According to the task, the value for n is 4, since we consider four dispersal risk parameters. Below we present the model equation which has been fitted by following Eq. (4.21):

$$Output(x) = lin\left(w_{1,\lambda}\mu_{NSF}^{p_1} + w_{2,\lambda}\mu_{ASR}^{p_2} + w_{3,\lambda}\mu_{VIA}^{p_3} + w_{4,\lambda}\mu_{LDD}^{p_4}\right). \tag{8.15}$$

The values $w_{1,\lambda}$ to $w_{4,\lambda}$ with respect to λ are the corresponding grades of importance weights of the parameters *NSF*, *ASR*, *VIA*, *LDD* respectively obtained as in Table 8.14, and μ_{NSF}, μ_{ASR}, μ_{VIA}, and μ_{LDD} are the membership functions of those parameters. The values p_1 to p_4 are the weights that have to be determined for each membership value of the corresponding parameter.
For Model III, membership functions used in models developed via Yager, Dombi, and Dubois operators (see Eqs. 8.5, 8.7, 8.8, and 8.12) have been considered. For Model IV, membership functions used in the Model I constructed via the Hamacher operator (see Eqs. 8.5, 8.9, 8.10, 8.11) have been considered.

8.9.2 MODEL III

Here we use the same data set which has been used in the previous models. This model consists of four plant categories which execute as per the below algorithm:

- vector p denotes the original numerical values of the four parameters NSF, ASR, VIA, and LDD of a particular plant species.
- vector q denotes the membership values corresponding to Eqs. (8.5), (8.12), (8.7), and (8.8) of the four parameters NSF, ASR, VIA, and LDD of a particular plant species.
- $lin()$ denotes the numerical-linguistic transformation functions.

Enter vector $p = [x_{NSF}, x_{ASR}, x_{VIA}, x_{LDD}]$
Enter vector $q = [\mu_{NSF}(x), \mu_{ASR_\gamma}(x), \mu_{VIA}(x), \mu_{LDD}(x)]$
If $ASR \leq 20000$ (**Category I**)
$Output(x)=lin\left(w_{1,0.9}\mu_{NSF}{}^{2}(x) + w_{2,0.9}\mu_{ASR}{}^{30}(x) + w_{3,0.9}\mu_{VIA}{}^{20}(x)\right.$
$\left.+w_{4,0.9}\mu_{LDD}{}^{40}(x)\right)$
Else if $ASR < 20000$ and $NSF \leq 200$ (**Category II**)
$Output(x)=lin\left(w_{1,0.9}\mu_{NSF}{}^{20}(x) + w_{2,0.9}\mu_{ASR}{}^{50}(x) + w_{3,0.9}\mu_{VIA}{}^{40}(x)\right.$
$\left.+w_{4,0.9}\mu_{LDD}{}^{40}(x)\right)$
Else if $ASR \geq 10^6$ (**Category III**)
$Output(x)=lin\left(w_{1,0.8}\mu_{NSF}{}^{120}(x) + w_{2,0.8}\mu_{ASR}{}^{20}(x) + w_{3,0.8}\mu_{VIA}{}^{40}(x)\right.$
$\left.+w_{4,0.8}\mu_{LDD}{}^{80}(x)\right)$
Else if $(ASR < 1000 \text{and } LDD > 5)$ or $(ASR \leq 1000 \text{and } LDD < 3)$ (**Category IV**)
$Output(x)=lin\left(w_{1,0.9}\mu_{SF}(x) + w_{2,0.9}\mu_{ASR}(x) + w_{3,0.9}\mu_{VIA}(x) + w_{4,0.9}\mu_{LDD}(x)\right)$
End
End
End
End

The risk evaluation in each category mentioned in the above algorithm is discussed below:

Category I: The species with $ASR \leq 20000$ belongs to this category. The values of p for the parameters NSF, ASR, VIA, and LDD are 2, 30, 20, 40 respectively. The value of parameter λ takes 0.9 for each parameter.

Category II: This group executes with the species having $ASR < 20000$ and $NSF \leq 200$. The values of p for the parameters NSF, ASR, VIA, and LDD are 20, 50, 40, 40 respectively. As in Category I, the value of λ takes 0.9 for each parameter.

Category III: The species with $ASR \geq 10^6$ belongs to this category. The p values take much higher values compared to the other categories such as 120, 20, 40, and 80 for NSF, ASR, VIA, LDD respectively. The value of λ takes 0.8 for each parameter.

Table 8.15

Test results - Model III Adapted from H.O.W. Peiris, S. Chakraverty, S.S.N. Perera, S.S.N and S.M.W. Ranwala, "Modelling Dispersal Risk of Invasive Alien Plant Species", In *Recent Advances in Applications of Computational and Fuzzy Mathematics*. (Springer 2018) 109-145.

Invasive Species	Model III Risk Level	NRA Risk Level
Alternanthera philoxeroides	*Very Low*	*Low*
Clidemia hirta	*High*	*High*
Miconia calvescens	*Very High*	*Extremely High*
Alstonia macrophylla	*Medium*	*Very High*
Annona glabra	*Very Low*	*Very Low*
Clusia rosea	*Medium*	*Medium*
Dillenia suffructicosa	*Very Low*	*Very Low*
Ageratina riparia	*Medium*	*High*
Mimosa invisa	*High*	*High*
Myroxylon balsamum	*Medium*	*Medium*
Tithonia diversiflora	*Medium*	*Low*
Mikania micrantha	*Medium*	*High*
Prosopis juliflora	*High*	*High*
Ulex europaeus	*High*	*Medium*
Mimosa pigra	*High*	*High*
Chromolaena odorata	*Medium*	*High*
Parthenium hysterophorus	*Medium*	*High*
Lantana camara	*Medium*	*Medium*
Imperata cylindrica	*High*	*High*

Category IV: The last category consists of species where $(ASR < 1000 \text{and} LDD > 5)$ or $(ASR \leq 1000 \text{and} LDD < 3)$. The p value for each parameter is 1 and λ is 0.9.

The model has been validated using the same set of species as in the previous models. The test results and validation results are represented in Tables 8.15 and 8.16.

8.9.3 MODEL IV

This model has been built upon the membership functions which were also used in the Model I–Hamacher. The algorithm for this model executes similarly to Model III.

The test and validation results are presented in Tables 8.17 and 8.18.

Table 8.16

Validation results - Model III. Adapted from H.O.W. Peiris, S. Chakraverty, S.S.N. Perera, S.S.N and S.M.W. Ranwala, "Modelling Dispersal Risk of Invasive Alien Plant Species", In *Recent Advances in Applications of Computational and Fuzzy Mathematics.* (Springer 2018) 109-145.

Category of Species	Species	Model III Risk Level	NRA Risk Level
Invasive	*Sphagneticola trilobata*	*Medium*	*Medium*
	Cuscuta campestris	*Medium*	*Medium*
	Pueraria montana	*Medium*	*Low*
Non-Invasive	*Cassia fistula*	*Very Low*	*Low*
	Cissus rotundifolia	*Very Low*	*Low*
	Hedychium gardnerianum	*Very Low*	*Very Low*
	Mangifera indica	*Very Low*	*Low*

Table 8.17

Test results - Model IV Adapted from H.O.W. Peiris, S. Chakraverty, S.S.N. Perera, S.S.N and S.M.W. Ranwala, "Modelling Dispersal Risk of Invasive Alien Plant Species", In *Recent Advances in Applications of Computational and Fuzzy Mathematics.* (Springer 2018) 109-145.

Invasive Species	Model IV Risk Level	NRA Risk Level
Alternanthera philoxeroides	*Low*	*Low*
Clidemia hirta	*High*	*High*
Miconia calvescens	*Very High*	*Extremely High*
Alstonia macrophylla	*Medium*	*Very High*
Annona glabra	*Very Low*	*Very Low*
Clusia rosea	*Medium*	*Medium*
Dillenia suffructicosa	*Very Low*	*Very Low*
Ageratina riparia	*Medium*	*High*
Mimosa invisa	*High*	*High*
Myroxylon balsamum	*Medium*	*Medium*
Tithonia diversiflora	*Low*	*Low*
Mikania micrantha	*Medium*	*High*
Prosopis juliflora	*High*	*High*
Ulex europaeus	*High*	*Medium*
Mimosa pigra	*High*	*High*
Chromolaena odorata	*Medium*	*High*
Parthenium hysterophorus	*Medium*	*High*
Lantana camara	*Medium*	*Medium*
Imperata cylindrica	*High*	*High*

Table 8.18

Validation results - Model IV Adapted from H.O.W. Peiris, S. Chakraverty, S.S.N. Perera, S.S.N and S.M.W. Ranwala, "Modelling Dispersal Risk of Invasive Alien Plant Species", In *Recent Advances in Applications of Computational and Fuzzy Mathematics*. (Springer 2018) 109-145.

Category of Species	Species	Model IV Risk Level	NRA Risk Level
Invasive	*Sphagneticola trilobata*	*Low*	*Medium*
	Cuscuta campestris	*Low*	*Medium*
	Pueraria montana	*Low*	*Low*
Non-Invasive	*Cassia fistula*	*Very Low*	*Low*
	Cissus rotundifolia	*Very Low*	*Low*
	Hedychium gardnerianum	*Very Low*	*Very Low*
	Mangifera indica	*Very Low*	*Low*

8.10 DISCUSSION – MODELS III AND IV

According to Table 8.15, one may see that risk levels of species *Alternanthera philoxeroides*, *Alstonia macrophylla*, *Ageratina riparia*, *Tithonia diversiflora*, *Mikania micrantha*, *Ulex europaeus*, *Chromolaena odorata*, and *Parthenium hysterophorus* are one/two level(s) behind/above the expected NRA levels.

In Model IV, among the species above mentioned, except for *Alternanthera philoxeroides* and *Tithonia diversiflora*, other species follow a similar pattern as in Model III (see Table 8.17).

According to the validation results of both models as presented in Tables 8.16 and 8.18, each non-invasive species takes risk level "*Very Low*" in both models which is a low level compared to NRA level, i.e. what we expect in reality. However, species of the invasive category takes a "*Medium*" risk level in Model III whereas the same group takes a "*Low*" risk level in Model IV which is the acceptable level.

Therefore, considering the test and validation results, it can be concluded that Model IV gives more significant results than Model III.

9 Evaluation of Invasion Risk Intervals with Quantitative and Qualitative Parameters

9.1 AN OVERVIEW

In the previous chapter we worked out key parameters of dispersal risk of invasive plant species. Despite the dispersal risk parameters, there are key parameters which can strongly influence invasiveness of plant species such as seed germination requirement level, vegetative reproduction strength, alleopathy property, etc. Unlike the quantitative parameters in the dispersal category, the above mentioned parameters are qualitative in nature. One cannot work with these parameters using the models given in Chapter 4 as they execute only with quantitative parameters. Therefore, in this work, methods presented in Chapter 5 based on multiple linear regression have been considered to evaluate the invasion risk of IAPS in the form of intervals.

Let us first discuss the risk factors/biological traits of IAPS which act as the model parameters in this work.

9.2 MODEL PARAMETERS

So far we have worked with dispersal risk factors. Despite these factors, the invasive potential of species can be recognized by the strength of vegetative reproduction and diversity of dispersal mechanism because these enable IAPS to occupy invaded habitat quickly and disperse with a far range (Ranwala, 2010). Moreover interfering competition based on alleopathy and physical defensive structures makes invasive pants more invasive. Therefore concerning these aspects, 12 biological traits/risk factors have been chosen including the dispersal factors as the model parameters. These have been selected from the National Risk Assessment (NRA) for alien invaders in Sri Lanka (Ranwala, 2010) and may be written as follows:

- Number of seeds per fruit (NSF),
- Annual seed production per m^2 (ASR),
- Viability of seeds (VIA),
- Long distance dispersal strength (LDD),
- Vegetative reproduction strength (VRS),
- Seed germination requirement level (SGL),
- Presence of physical defensive structures (PDS),
- Formation dense tickets (FDS),
- Potential to be spread by human activities (HA),

- Role of natural and man-made disturbances (*NMD*),
- Alleopathic property (*AP*),
- Existence of invasive races (*IR*).

The above mentioned factors are qualitative in nature except for the first four factors (dispersal factors). The data set of 33 known invasive alien species and a few non-invasive species has been obtained by the invasive specialists group attached to the Ministry of Environment and Renewable Resources, Sri Lanka. Here data of some invasive species and non-invasive species have been used for model validation. It contained single-valued observations of 12 parameters and invasion risk scores which are obtained from the NRA procedure using the checklist prepared and accepted by the Ministry of Environment and Renewable Resources, Sri Lanka, after broad stakeholder participation and discussion (Ranwala, 2010).

Below we present the model which evaluates the invasion risk intervals of IAPS based on the 12 parameters mentioned above.

9.3 MODEL – INVASION RISK INTERVALS OF IAPS

Here the invasion risk interval Inv_R of a particular alien plant species is assumed to depend on 12 parameters: *NSF*, *ASR*, *VIA*, *LDD*, *VRS*, *SGL*, *PDS*, *FDS*, *HA*, *NMD*, *AP*, and *IR*. Accordingly, a linear relation is assumed for Inv_R as

$$\begin{aligned} Inv_R = \tilde{\theta_0} + \tilde{\theta_1}(NSF) + \tilde{\theta_2}(ASR) + \tilde{\theta_3}(VIA) + \tilde{\theta_4}(LDD) + \tilde{\theta_5}(VRS) + \tilde{\theta_6}(SGL) \\ + \tilde{\theta_7}(PDS) + \tilde{\theta_8}(FDS) + \tilde{\theta_9}(HA) + \tilde{\theta_{10}}(NMD) + \tilde{\theta_{11}}(AP) + \tilde{\theta_{12}}(IR). \end{aligned} \tag{9.1}$$

Here all the components such as Inv_R, input values of the parameters, and coefficients (θ)are in the form of intervals. To develop the model, the interval coefficients have to be approximated by using the methods given in Chapter 5.

First the interval-valued input data matrix $\ddot{X}$ has to be constructed since the data of each parameter is in single-valued points. Here NRA scores represent Inv_R in the process of approximating interval coefficients.
In the following subsection we explain the procedure that has been followed to convert the single-valued data into interval data.

9.3.1 FORMULATION OF INTERVAL-VALUED DATA

One may note that the parameters such as VRS, SGR, HA, NMD are qualitative parameters described by linguistic labels and each label is assigned a numerical value. For example, the strength of vegetative reproduction is explained by five linguistic labels such as *VeryLow*, *Low*, *Medium*, *High*, *VeryHigh* and a numerical score for each label is assigned as 1, 2, 3, 4, 5 respectively.

In the process of interval data formulation, scores assigned to the linguistic labels of qualitative parameters and real data of quantitative parameters have been converted into intervals by performing width adjustments considering those values as centers. Here, the nature of each parameter is assumed to form interval-valued data and to keep the essence of experts' opinions for risk scores. One may note that data for the parameters *SGR*, *PDS*, *AP*, and *IR* are in the form of yes/no answers. In reality the term yes/no is exact, with no need of converting into intervals. Therefore the scores assigned to the yes/no answers in the NRA, i.e. 1 and 0, have been considered as the data for those parameters. It has been found that the appropriate widths from center to end point of parameters and NRA score are ± 0.4,± 5 respectively. It may be noted that the lower and upper boundaries of interval-valued data are all non-negative values.

After evaluating the interval input matrix, the next step is to approximate the interval coefficients ($\tilde{\theta}_0$) in Eq. (9.1). For that three approaches mentioned in Chapter 5 have been used and three models have been developed. In the following section, we discuss each model in detail.

9.4 MODEL I

Here the model equation is the same as Eq. (9.1) and the interval coefficients are approximated by following Approach I in Chapter 5. First the initial solution has to be approximated by taking the original data without converting into intervals. Using the least squares method as stated in Subsection 5.2.2., point solutions for the coefficients can be easily found. After that, following the steps mentioned in Approach I, the interval coefficients can be approximated. The approximated interval coefficients and the initial point solution which is considered as the center values of the coefficients are presented in Table 9.1.

9.5 MODEL II

Here we develop the model by assessing the interval coefficients in Eq. (9.1) by following Approach II in Chapter 5. In order to apply the method, the following conditions need to be satisfied:

- Initial point estimations of coefficients have to be non-negative values.
  This condition can be verified from the second column of Table 9.1. The values in that column represent the initial point solutions of the coefficients and all are non-negative values.
- Lower and upper boundary values of the parameters and expected output should be non-negative values.
  Here the expected output is NRA scores. All the parameter values along with NRA scores are non-negative values. After converting the parameters and NRA scores into intervals, it has been observed that the lower and upper boundary points are non-negative.

The coefficients of Eq.(9.1) have been approximated and are presented in Table. 9.2.

Table 9.1

Interval estimates for coefficients of Model I Adapted from H.O.W. Peiris, S. Chakraverty, S.S.N. Perera, S.S.N and S.M.W. Ranwala, "Novel interval multiple linear regression model to assess the risk of invasive alien plant species," *Journal of Science,* Southeastern University of Sri Lanka 9, no(1) (2018): 12-30.

Coefficient	Center Value of Coefficient	Interval Estimates-Model I
θ_0	11.07908	$[11.07323, 11.08492]$
θ_1	0.0183830220805151	$[0.01837, 0.01840]$
θ_2	$5.40300982218465 \times 10^{-6}$	$[5.396 \times 10^{-6}, 5.409 \times 10^{-6}]$
θ_3	0.0155453826052045	$[0.01553, 0.01556]$
θ_4	0.0385944399652945	$[0.03855, 0.03864]$
θ_5	2.14988781032716	$[2.14781, 2.15196]$
θ_6	2.69472395393735	$[2.69187, 2.69758]$
θ_7	2.43003869540741	$[2.42756, 2.43251]$
θ_8	1.9773208835343	$[1.97549, 1.97915]$
θ_9	3.07114393755476	$[3.06775, 3.07454]$
θ_{10}	1.34943726439762	$[1.34851, 1.35037]$
θ_{11}	2.42486246924465	$[2.42239, 2.42733]$
θ_{12}	2.51326594838486	$[2.51067, 2.51586]$

Table 9.2

Interval estimates for coefficients of Model II Adapted from H.O.W. Peiris, S. Chakraverty, S.S.N. Perera, S.S.N and S.M.W. Ranwala, "Novel interval multiple linear regression model to assess the risk of invasive alien plant species," *Journal of Science,* Southeastern University of Sri Lanka 9, no(1) (2018): 12-30.

Coefficient	Center Value of Coefficient	Interval Estimates-Model II
θ_0	11.07908	$[9.81417, 12.34399]$
θ_1	0.0183830220805151	$[0.0183830220805139, 0.0183830220805163]$
θ_2	$5.40300982218465 \times 10^{-6}$	$[5.40300982218348 \times 10^{-6}, 5.40300982218582 \times 10^{-6}]$
θ_3	0.0155453826052045	$[0.0155453826052041, 0.0155453826052049]$
θ_4	0.0385944399652945	$[0.0385944399652919, 0.0385944399652971]$
θ_5	2.14988781032716	$[2.14988781032714, 2.14988781032718]$
θ_6	2.69472395393735	$[2.69472395393733, 2.69472395393737]$
θ_7	2.43003869540741	$[2.43003869540739, 2.43003869540743]$
θ_8	1.9773208835343	$[1.97732088353429, 1.97732088353431]$
θ_9	3.07114393755476	$[3.07114393755475, 3.07114393755477]$
θ_{10}	1.34943726439762	$[1.3494372643976, 1.34943726439764]$
θ_{11}	2.42486246924465	$[2.42486246924464, 2.42486246924466]$
θ_{12}	2.51326594838486	$[2.51326594838485, 2.51326594838487]$

9.6 COMPARISON OF QUALITY OF MODELS I AND II

The average accuracy ratio as stated in Eq. (5.17) has been used to compare the overall quality of Models I and II. Table 9.3 summarizes the accuracy ratio of each model. Figures 9.1 and 9.2 show a graphical comparison among expected lower and

Table 9.3
Quality comparison between Model I and Model II Adapted from H.O.W. Peiris, S. Chakraverty, S.S.N. Perera, S.S.N and S.M.W. Ranwala, "Novel interval multiple linear regression model to assess the risk of invasive alien plant species," *Journal of Science,* Southeastern University of Sri Lanka 9, no(1) (2018): 12-30.

Model No	Average Accuracy Ratio
Model I	0.65779
Model II	0.727641

upper risk boundaries with approximated lower, and upper risk boundaries that have been obtained from both models for 28 invasive plant species in the data set. In these two figures, symbol $'*'$, $'\times'$, $'\boxdot'$ and $'\triangle'$ represent expected lower, expected upper, approximated lower, and approximated upper boundary of risk respectively.

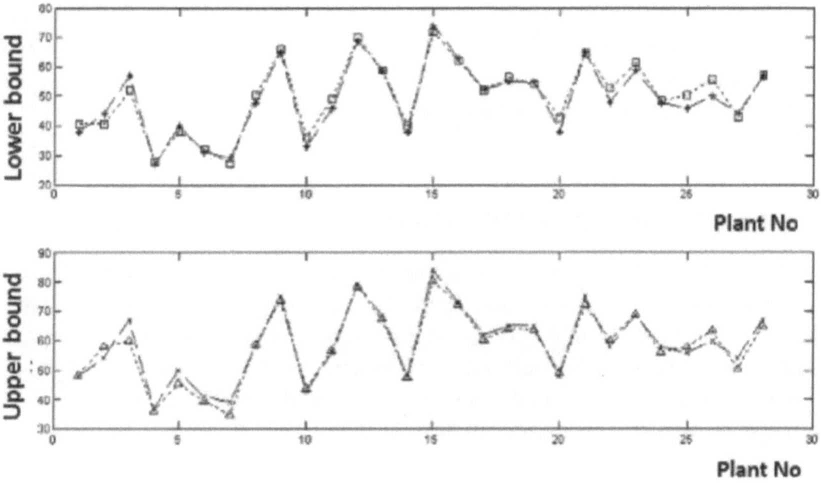

Figure 9.1 Comparison of expected risk boundaries with approximated risk boundaries from Model I. (Adapted from H.O.W. Peiris, S. Chakraverty, S.S.N. Perera, S.S.N and S.M.W. Ranwala, "Novel interval multiple linear regression model to assess the risk of invasive alien plant species," *Journal of Science,* Southeastern University of Sri Lanka 9, no(1) (2018): 12-30.)

9.7 VALIDATION RESULTS

The two models have been validated using known invasive and non-invasive species. The results are presented in Table 9.4.

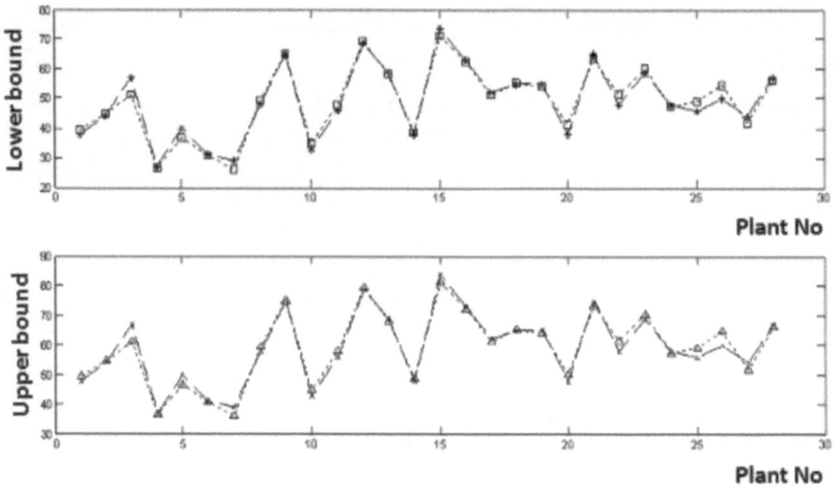

Figure 9.2 Comparison of expected risk boundaries with approximated risk boundaries from Model II. (Adapted from H.O.W. Peiris, S. Chakraverty, S.S.N. Perera, S.S.N and S.M.W. Ranwala, "Novel interval multiple linear regression model to assess the risk of invasive alien plant species," *Journal of Science,* Southeastern University of Sri Lanka 9, no(1) (2018): 12-30.)

9.8 DISCUSSION - MODELS I AND II

According to Table 9.4., it can be seen that both models reflect the invasiveness of plant species more or less the same way that a group of experts have given a rank in the national risk assessment procedure. It is obvious that the known invasive species have obtained a higher risk value compared to non-invasive species. Therefore, it is clear that the model has potential to serve as a tool which could screen the invasiveness of species before considering for a country.

The $\varepsilon-$ inflation process in Model I computationally finds the interval regression coefficients with reasonable quality. If the decision maker is satisfied with the value of the ratio estimation, then the current ε value will be the width of the interval coefficient. However, it is not easy to find which ε value fits better with a particular coefficient. According to Tables 9.1 and 9.2, the boundaries of the coefficients of Model I and II are positively related to the invasion risk. In this kind of situation Model II can be easily applied. On the other hand accuracy ratios of the models in Table 9.3 reveal how well the expected outcome intersects with approximated outcome. It can clearly be seen that Model II reflects a higher level of intersection compared to Model I and Figures 9.1 and 9.2 confirm that the expected outcome and approximated outcome of Model II are closer than Model I.

Here we consider that the model outcome is a better interpretation of the risk of a species if the NRA score is within that output interval. For example, if we consider

Table 9.4

Validation results - Model I & II Adapted from H.O.W. Peiris, S. Chakraverty, S.S.N. Perera, S.S.N and S.M.W. Ranwala, "Novel interval multiple linear regression model to assess the risk of invasive alien plant species," *Journal of Science*, Southeastern University of Sri Lanka 9, no(1) (2018): 12-30.

Category of Species	Species	NRA score(%)	Approximated Risk - Model I	Approximated Risk - Model II
Invasive	*Austroeupator iuminulifolium*	62	[56.333, 63.91]	[55.118, 65.118]
	Panicum maximum	66	[64.495, 72.091]	[63.29, 73.289]
	Cuscuta campestris	60	[55.506, 63.082]	[54.29, 64.29]
	Pueraria montana	55	[54.777, 62.352]	[53.561, 63.561]
	Acacia mearnsii	64	[57.781, 65.364]	[56.5689, 66.569]
Non-Invasive	*Myrica faya*	36	[34.282, 41.817]	[33.045, 43.045]
	Cassia fistula	32	[32.24, 39.773]	[31.003, 41.003]
	Cissus rotundifolia	32	[31.26, 38.788]	[30.02, 40.02]
	Hedychium gardnerianum	32	[33.888, 41.421]	[32.651, 42.650]
	Magnefera indica	32	[32.287, 39.82]	[31.05, 41.05]

the non-invasive species *Cassia fistula*, which has an NRA score of 32, this risk score is out of the boundaries of the predicted risk interval in Model I, but it is within the boundaries of the risk interval in Model II. The results are similar for non-invasive species *Magnefera indica*. On the other hand, the species *Hedychium gardnerianum* with NRA score of 32 is out of the boundaries of estimated risk intervals in both models. But the lower boundary of Model II is closer to the NRA score than the lower boundary of Model I. According to the discussion, Model II can be considered as the tool which gives a significant outcome as to the risks of IAPS.

9.9 MODEL III

The model that we discuss in this section is based on the Approach III in Chapter 5. Here the interval coefficients are approximated using the method stated in Approach III. As in Model II, the same set of conditions may be applicable for this model.

Here we have formulated several interval input data sets by performing width adjustments in crisp data. In that task, the nature of each parameter is assumed to form interval-valued data and keep the essence of experts' opinions for risk scores. Table 9.5 presents the interval input data sets by indicating the spread of parameters and NRA scores. It may be noted that the lower and upper boundaries of interval-valued data are all non-negative values.

Afterwards, interval coefficients of Eq. (9.1) have been estimated for each interval input data set by following Approach III.

Table 9.5

Width adjustments of parameters Adapted from H.O.W. Peiris, S. Chakraverty, S.S.N. Perera, S.S.N and S.M.W. Ranwala, "Development of a risk assessment mathematical model to evaluate invasion risk of invasive alien species using interval multivariate linear regression," *British Journal of Applied Mathematics and Technology* 16, no(1) (2016): 1-11.

Data set	Spread from center of interval-valued input data of parameters	Spread from center of interval-valued risk scores	
		Left	Right
1	±0.4	3	3
2	±0.5	4	4
3	±0.4	5	5
4	±0.5	4	2

Table 9.6

Interval estimates of coefficients from data set 1 Adapted from H.O.W. Peiris, S. Chakraverty, S.S.N. Perera, S.S.N and S.M.W. Ranwala, "Development of a risk assessment mathematical model to evaluate envasion risk of invasive alien species using interval multivariate linear regression," *British Journal of Applied Mathematics and Technology* 16, no(1) (2016): 1-11.

Coefficient	Estimates from data set 1
θ_0	$[10.344, 11.814]$
θ_1	$[0.0183830220805138, 0.0183830220805161]$
θ_2	$[5.40300982218432 \times 10^{-6}, 5.40300982218483 \times 10^{-6}]$
θ_3	$[0.015545382605203, 0.015545382605205]$
θ_4	$[0.03859443996525, 0.03859443996527]$
θ_5	$[2.1499, 2.1499]]$
θ_6	$[2.69472395393732, 2.69472395393747]$
θ_7	$[2.43003869540741, 2.43003869540748]$
θ_8	$[1.9773, 1.9773]$
θ_9	$[3.07114393755477, 3.07114393755487]$
θ_{10}	$[1.34943726439752, 1.34943726439766]$
θ_{11}	$[2.42486246924464, 2.42486246924466]$
θ_{12}	$[2.51326594838487, 2.51326594838489]$

Tables 9.6 to 9.9 summarize the interval estimations of coefficients of the model assessed in each interval-input data set presented in Table 9.5.

9.10 QUALITY MEASUREMENT OF MODEL III

To measure the overall quality of the model, we have used average accuracy ratio as defined in Eq.(5.17). Table 9.10 summarizes the average accuracy ratios of the model incorporated with each data set given in Table 9.5. Figures 9.3 to 9.6 show graphical comparison with expected and approximated lower and upper risk boundaries for

Table 9.7

Interval estimates of coefficients from data set 2 Adapted from H.O.W. Peiris, S. Chakraverty, S.S.N. Perera, S.S.N and S.M.W. Ranwala, "Development of a risk assessment mathematical model to evaluate invasion risk of invasive alien species using interval multivariate linear regression," *British Journal of Applied Mathematics and Technology* 16, no(1) (2016): 1-11.

Coefficient	Estimates from data set 2
θ_0	$[10.3943, 11.745]$
θ_1	$[0.0183106933740867, 0.0183798813271578]$
θ_2	$[5.40142448033662 \times 10^{-6}, 5.40644416509869 \times 10^{-6}]$
θ_3	$[0.0155469931613369, 0.0155539824393268]$
θ_4	$[0.0391974418649532, 0.0414273985753972]$
θ_5	$[2.15044615933592, 2.15203670731178]]$
θ_6	$[2.69230683186782, 2.69508724936171]$
θ_7	$[2.42696169733635, 2.43008821070789]$
θ_8	$[1.97708136819981, 1.97821672074128]$
θ_9	$[3.0650823787639, 3.07073314364203]$
θ_{10}	$[1.348870752105111.35368206335038]$
θ_{11}	$[2.42463624793798, 2.42513167844029]$
θ_{12}	$[2.51340075543958, 2.51512964449058]$

Table 9.8

Interval estimates of coefficients from data set 3 Adapted from H.O.W. Peiris, S. Chakraverty, S.S.N. Perera, S.S.N and S.M.W. Ranwala, "Development of a risk assessment mathematical model to evaluate invasion risk of invasive alien species using interval multivariate linear regression," *British Journal of Applied Mathematics and Technology* 16, no(1) (2016): 1-11.

Coefficient	Estimates from data set 3
θ_0	$[9.65394, 12.34399]$
θ_1	$[0.0183830220805151, 0.01886]$
θ_2	$[5.40300982218465 \times 10^{-6}, 5.81 \times 10^{-6}]$
θ_3	$[0.0155453826052045, 0.01555]$
θ_4	$[0.0385944399652945, 0.07717]$
θ_5	$[2.14988781032716, 2.19395]]$
θ_6	$[2.69472395393735, 2.71626]$
θ_7	$[2.43003869540741, 2.44489]$
θ_8	$[1.93169, 1.9773208835343]$
θ_9	$[3.0212, 3.07114393755476]$
θ_{10}	$[1.27264, 1.34943726439764]$
θ_{11}	$[2.42486246924465, 2.47308]$
θ_{12}	$[2.51326594838486, 2.5336]$

28 invasive plant species in the data set. In these figures, symbol $'*'$, $\times$, $'\dagger'$ and $'\triangle'$ represent expected lower, expected upper, approximated lower, and approximated upper boundary of risk respectively.

Table 9.9

Interval estimates of coefficients from data set 4 Adapted from H.O.W. Peiris, S. Chakraverty, S.S.N. Perera, S.S.N and S.M.W. Ranwala, "Development of a risk assessment mathematical model to evaluate invasion risk of invasive alien species using interval multivariate linear regression," *British Journal of Applied Mathematics and Technology* 16, no(1) (2016): 1-11.

Coefficient	Estimates from data set 4
θ_0	$[8.3943, 11.745]$
θ_1	$[0.0183106933740868, 0.0183798813271578]$
θ_2	$[5.40142448033665 \times 10^{-6}, 5.40644416509869 \times 10^{-6}]$
θ_3	$[0.0155469931613369, 0.0155539824393267]$
θ_4	$[0.0391974418649532, 0.0414273985753901]$
θ_5	$[2.15044615933592, 2.15203670731174]]$
θ_6	$[2.69230683186785, 2.69508724936171]$
θ_7	$[2.42696169733638, 2.43008821070789]$
θ_8	$[1.97708136819981, 1.97821672074126]$
θ_9	$[3.06508237876395, 3.07073314364203]$
θ_{10}	$[1.34887075210511, 1.3536820633504]$
θ_{11}	$[2.42463624793801, 2.42513167844029]$
θ_{12}	$[2.51340075543958, 2.51512964449059]$

9.11 VALIDATION - MODEL III

The model with four different interval coefficients has been validated with the same data set as was used in the validation of Models I and II. The validation results are summarized in Tables 9.11 and 9.12.

9.12 DISCUSSION - MODEL III

According to Tables 9.6 to 9.9, it can be seen that the width spread of estimated interval coefficients is changing with respect to interval input data set. One may

Table 9.10

Quality comparison Model III Adapted from H.O.W. Peiris, S. Chakraverty, S.S.N. Perera, S.S.N and S.M.W. Ranwala, "Development of a risk assessment mathematical model to evaluate invasion risk of invasive alien species using interval multivariate linear regression," *British Journal of Applied Mathematics and Technology* 16, no(1) (2016): 1-11.

Data set	Average accuracy ratio
1	0.60306
2	0.679851
3	0.730852
4	0.603521

Table 9.11

Validation results I - Model III Adapted from H.O.W. Peiris, S. Chakraverty, S.S.N. Perera, S.S.N and S.M.W. Ranwala, "Development of a Risk Assessment Mathematical Model to Evaluate Invasion Risk of Invasive Alien Species Using Interval Multivariate Linear Regression," *British Journal of Applied Mathematics and Technology* 16, no(1) (2016): 1-11.

Category of Species	Species	NRA score(%)	Approximated Risk Data set 1	Approximated Risk Data set 2
Invasive	*Austroeupator iuminulifolium*	62	[55.64757, 64.58792]	[54.75459, 65.50676]
	Panicum maximum	66	[63.8192, 72.75955]	[62.9152, 73.66792]
	Cuscuta campestris	60	[54.81983, 63.76019]	[53.92563, 64.67496]
	Pueraria montana	55	[54.0905, 63.03085]	[53.17714, 63.93692]
	Acacia mearnsii	64	[57.09867, 66.03903]	[56.1979, 66.94356]
	Myrica faya	36	[33.57519, 42.51555]	[32.67957, 43.40609]
Non-Invasive	*Cassia fistula*	32	[31.53293, 40.47328]	[30.62525, 41.36012]
	Cissus rotundifolia	32	[30.55005, 39.49041]	[29.65617, 40.38129]
	Hedychium gardnerianum	32	[33.18063, 42.12099]	[32.28433, 43.01214]
	Magnefera indica	32	[31.57978, 40.52013]	[30.67473, 41.41821]

Table 9.12

Validation results II - Model III Adapted from H.O.W. Peiris, S. Chakraverty, S.S.N. Perera, S.S.N and S.M.W. Ranwala, "Development of a risk assessment mathematical model to evaluate invasion risk of invasive alien species using interval multivariate linear regression," *British Journal of Applied Mathematics and Technology* 16, no(1) (2016): 1-11.

Category of Species	Species	NRA score(%)	Approximated Risk Data set 3	Approximated Risk Data set 4
Invasive	*Austroeupator iuminulifolium*	62	[55.11774, 65.11774]	[52.75459, 65.50676]
	Panicum maximum	66	[63.28938, 73.28938]	[60.9152, 73.66792]
	Cuscuta campestris	60	[54.29001, 64.29001]	[51.92563, 64.67496]
	Pueraria montana	55	[53.56068, 63.56068]	[51.17714, 63.93692]
	Acacia mearnsii	64	[56.56885, 66.56885]	[54.1979, 66.94356]
	Myrica faya	36	[33.04537, 43.04537]	[30.67957, 43.40609]
Non-Invasive	*Cassia fistula*	32	[31.00311, 41.00311]	[28.62525, 41.36012]
	Cissus rotundifolia	32	[30.02023, 40.02023]	[27.65617, 40.38129]
	Hedychium gardnerianum	32	[32.65081, 42.65081]	[30.28433, 43.01214]
	Magnefera indica	32	[31.04996, 41.04996]	[28.67473, 41.41821]

observe that the largest width spread forms from data set 3. If we analyze the quality measurement of the model with respect to data sets, it can be seen that the average accuracy ratio of 0.730852 along with data set 3 is the highest where data set 1 gives the lowest value which is 0.60306. The variations among average accuracy ratios can be seen from Figs. 9.3 to 9.6 which depict the overlaps between estimated and

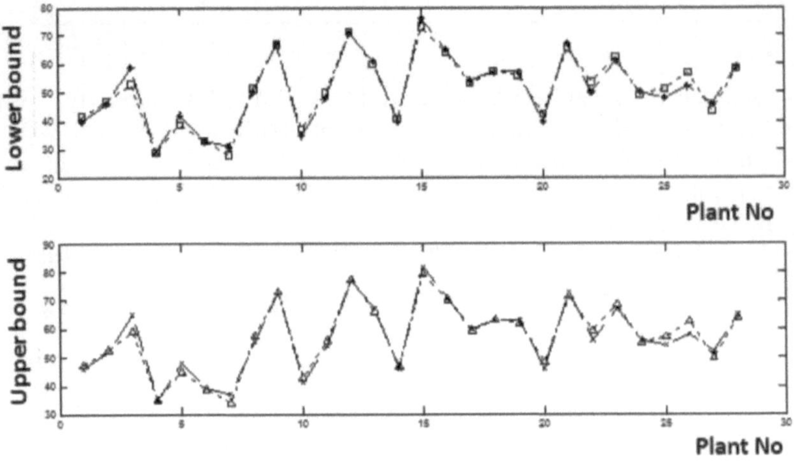

Figure 9.3 Interval estimations of risk of IAPS from data set 1. (Adapted from H.O.W. Peiris, S. Chakraverty, S.S.N. Perera, S.S.N and S.M.W. Ranwala, "Development of a risk assessment mathematical model to evaluate invasion risk of invasive alien species using interval multivariate linear regression," *British Journal of Applied Mathematics and Technology* 16, no(1) (2016): 1-11.)

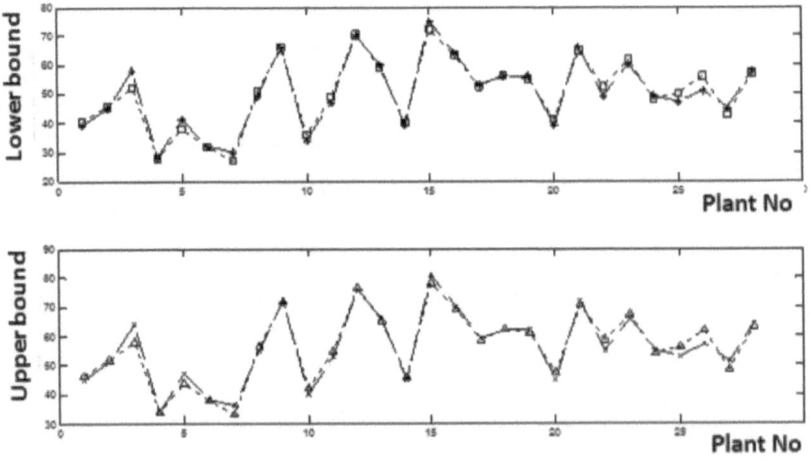

Figure 9.4 Interval estimations of risk of IAPS from data set 2. (Adapted from H.O.W. Peiris, S. Chakraverty, S.S.N. Perera, S.S.N and S.M.W. Ranwala, "Development of a risk assessment mathematical model to evaluate invasion risk of invasive alien species using interval multivariate linear regression," *British Journal of Applied Mathematics and Technology* 16, no(1) (2016): 1-11.)

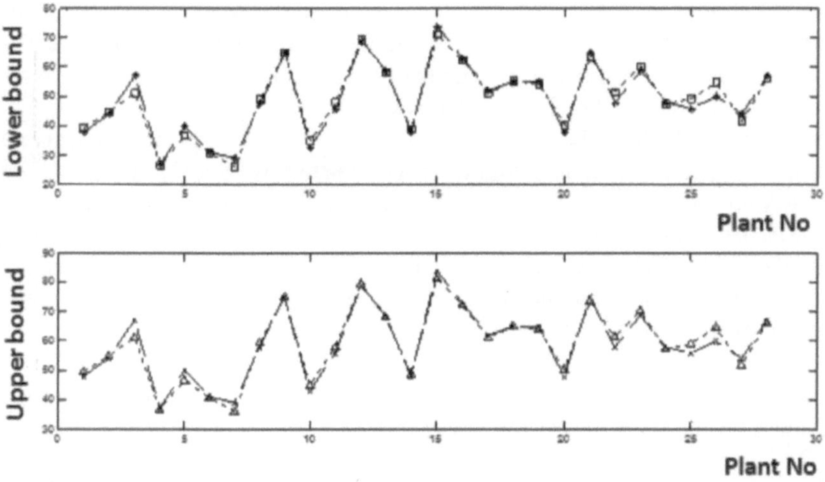

Figure 9.5 Interval estimations of risk of IAPS from data set 3. (Adapted from H.O.W. Peiris, S. Chakraverty, S.S.N. Perera, S.S.N and S.M.W. Ranwala, "Development of a risk assessment mathematical model to evaluate invasion risk of invasive alien species using interval multivariate linear regression," *British Journal of Applied Mathematics and Technology* 16, no(1) (2016): 1-11.)

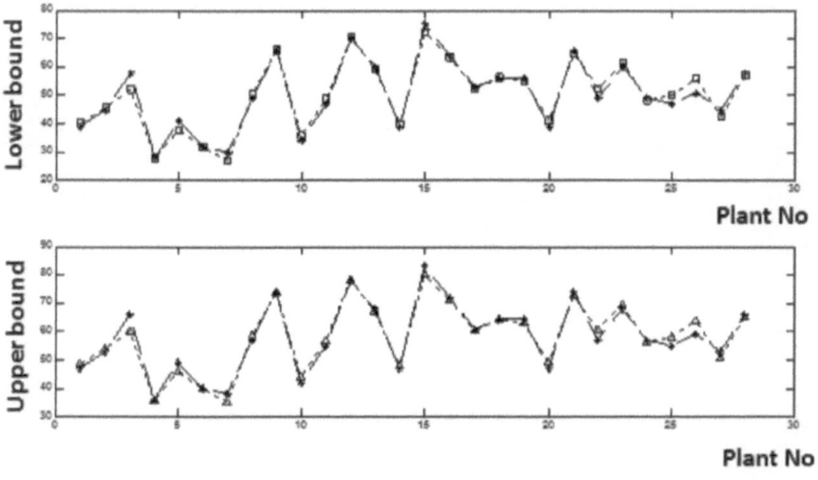

Figure 9.6 Interval estimations of risk of IAPS from data set 4. (Adapted from H.O.W. Peiris, S. Chakraverty, S.S.N. Perera, S.S.N and S.M.W. Ranwala, "Development of a risk assessment mathematical model to evaluate invasion risk of invasive alien species using interval multivariate linear regression," *British Journal of Applied Mathematics and Technology* 16, no(1) (2016): 1-11.)

expected risk intervals. If one may compare the model with data set 3 with Model II, it can be seen that the difference between the accuracy ratios of both models is negligible. Therefore one may conclude that these two models developed with two different procedures function in a similar way.

According to Tables 9.11 and 9.12, one may see that the expected risk score of each invasive plant species is within the approximated risk interval. It can be seen that NRA scores of species *Austroeupatorium inulifolium*, *Cuscuta campestris*, and *Acacia mearnsii* stand much closer to the upper boundary of the approximated level where the NRA of *Panicum maximum* and *Pueraria montana* stand closer to the lower boundary of the approximated level. This pattern can be seen in each approximated level via data set. On the other hand, NRA of non-invasive species stand closer to the lower boundary of the approximated interval except the species *Hedychium gardnerianum*. This result tallies with what we expect in reality, that is lower risk values for the non-invasive species.

The lower boundaries of estimated risk intervals from input data sets 2 and 3 are closer to NRA score than the lower boundary of input data set 1. Therefore comparing the quality and validation results, Model III with data set 3 and Model II should be incorporated for better prediction of the risk of invasive alien species.

10 Evaluation of Invasion Risk Levels with Symmetrically Distributed Linguistic Terms Sets

10.1 AN OVERVIEW

Here we apply the methods mentioned in Chapter 6 to develop risk models to assess the invasion risk of IAPS. As stated in Chapter 9, except for the dispersal-related parameters, other parameters which describe the invasiveness of IAPS are in qualitative form. As in Chapter 8, assessing dispersal risk, the final outputs of the models are also in qualitative form. Therefore, the methods in Chapter 6 can be applied to develop intended models. The same data set which has been used in the previous chapters is considered for this chapter.

In the first half of the work, we present three models which are developed through fuzzy linguistic operators: *LOWA*, *LWA*, and *MLIOWA*. The model with *LOWA* is accompanied by non-weighted parameters, i.e. the equally important aspect of parameters. And the other two models are based on weighted parameters such that the importance of parameters is also considered.

In the second half, two models are presented by changing the linguistic approach to a fuzzy 2-tuple approach. The developed models are based on the equal/unequal importance of the parameters.

Finally, we present a model which is developed based on the approximation method as stated in Chapter 6. Here the weights of the parameters are obtained using a linguistic scale which is different from the scale used for gathering the performance values of the parameters.

All six models have been validated using the same data set used in the validation in Chapter 9.

In the next section, we present the models developed via fuzzy linguistic operators.

10.2 MODELS – FUZZY LINGUISTIC OPERATORS

As we mentioned earlier the model that we are going to develop here is based on the model structures presented in Chapter 6. One may observe that, as depicted in

Figures 6.3 to 6.5, the first two steps are common in each structure. These steps are identifying the parameters and setting up the appropriate linguistics scale. We discuss these two steps in detail below.

10.2.1 IDENTIFYING PARAMETERS

Here we concern ourselves with evaluating overall invasion risk that emerges from biological traits of IAPS as was done in Chapter 9. As in Chapter 9, we consider the same set of parameters except presence of Physical defensive structures (*PDS*), Formation dense tickets (*FDS*), Alleopathic property (*AP*), Existence of invasive races (*IR*). The reason is these parameters have performances values with yes/no answers.

Therefore the following eight parameters have been selected as the model parameters:

- Number of seeds per fruit (*NSF*)
- Annual seed production per $m^2(ASR)$
- Viability of seeds (in months) (*VIA*)
- Long distance dispersal strength (*LDD*)
- Vegetative reproduction strength (*VRS*)
- Seed germination requirement level (*SGL*)
- Potential to be spread by human activities (*HA*)
- Role of natural and man-made disturbances (*NMD*)

Further, these parameters have been categorized into main and sub parameters. This task has been carried out by following the plant science experts' opinions (see Table 10.1). According to Table 10.1, it can be seen that there are four main parameters labeled as *DIS*, *VRS*, *SGL*, and *MIS*. There are four sub parameters for *DIS* and two sub parameters for *MIS*. And there are no sub parameters accompanied by the remaining main parameters.

Next we discuss the procedure to set up the appropriate linguistic terms set.

10.2.2 SELECTION PROCEDURE OF LINGUISTIC TERM SET

When selecting the appropriate linguistic terms set, performance values of the parameters and the nature of the output have to be considered. In Chapter 8, dispersal risk of IAPS has been formed in risk levels based on a uniform and symmetrical distributed linguistic terms set having seven labels (see Figure 4.4). Since other main/sub parameters are in qualitative form and the output is to be generated as risk levels, the same linguistic terms set as depicted in Figure 4.4 has been chosen.

Thus cardinality of the linguistic terms set is seven and the semantics of labels are presented as follows:

Table 10.1
Main and sub parameters of invasiveness

Main parameter	Sub parameter
Dispersal (*DIS*)	Number of seeds per fruit (*NSF*)
	Annual seed production per m^2(*ASR*)
	Viability of seeds (in months) (*VIA*)
	Long distance dispersal strength (*LDD*)
Vegetative reproduction strength (*VRS*)	-
Seed germination requirement level (*SGL*)	-
Man's influence on spreading (*MIS*)	Potential to be spread by human activities (*HA*)
	Role of natural and man-made disturbances (*NMD*)

$s_0 = U = Unlikely = (0,0,0.16)$
$s_1 = VL = Very\ Low = (0,0.16,0.34)$
$s_2 = L = Low = (0.16,0.34,0.5)$
$s_3 = M = Medium = (0.34,0.5,0.66)$
$s_4 = H = High = (0.5,0.66,0.84)$
$s_5 = VH = (0.66,0.84,1)$
$s_6 = EH = Extremely\ High = (0.84,1,1)$

These linguistic terms are placed symmetrically around the middle term which represents the assessment of "approximately 0.5" and satisfies the properties negation, maximization, minimization as defined in Subsection 6.2.1.

This linguistic terms set is used to evaluate the performance values of parameters and to generate the risk levels as the output.

In the next section, we discuss the model developed based on the *LOWA* operator.

10.2.3 MODEL I – *LOWA*

This model is developed by following the model structure as depicted in Figure 6.3. Since the operator is *LOWA*, the parameters have been considered as equally important towards invasiveness. As we have already covered the first two steps, let us begin with the third step.

Evaluation of performance values of the parameters One may note that, in Chapter 8, the dispersal risk model has been constructed with four quantitative parameters which are the sub parameters as in Table 10.1. Therefore, we have gathered performance values of IAPS in the data set directly from Model I–Hamacher. It reveals that the sub parameters of *DIS* do not have to be incorporated in the aggregation process in this model and only the performance values of *DIS* are needed. The performance values of the other main/sub parameters have been obtained based on the chosen linguistic terms set.

Aggregation using *LOWA* As we mentioned in Chapter 6, the *LOWA* operates with linguistic quantifiers. However, there is no specific procedure to choose the appropriate quantifier in the given context. Therefore, we have chosen three quantifiers such as Mean, Most, At least half.

The aggregation procedure runs through two steps. The first step is to aggregate the sub parameters of a main parameter and the other one is to aggregate the main parameters.

First of all let us calculate the weights generated from each quantifier in the aggregation of four main parameters as follows:

- Weights from Mean quantifier:
  As to Eq. (6.2), n is four since there are only four main parameters.
  Thus $w_1 = w_2 = w_3 = w_4 = \frac{1}{4} = 0.25$
  Hence the weighing vector is $[0.25, 0.25, 0.25, 0.25]$.
- Weights from Most quantifier:
  The parameter of the qualifier is $(0.3, 0.8)$.
  By applying Eq. (6.1) we get

$$w_1 = Q\left(\frac{1}{4}\right) - Q\left(\frac{0}{4}\right) = Q(0.25) = 0,$$

$$w_2 = Q\left(\frac{2}{4}\right) - Q\left(\frac{1}{4}\right) = Q(0.5) - Q(0.25) = \frac{0.5-0.3}{0.8-0.3} - 0 = 0.4,$$

$$w_3 = Q\left(\frac{3}{4}\right) - Q\left(\frac{2}{4}\right) = \frac{0.75-0.3}{0.8-0.3} - Q(0.5) = 0.9 - 0.4 = 0.5,$$

$$\text{and } w_4 = Q(1) - Q\left(\frac{3}{4}\right) = 1 - Q(0.75) = 1 - 0.9 = 0.1.$$

  Hence the weighting vector is $w = [0, 0.4, 0.5, 0.1]$.
- Weights from At Least half quantifier:
  The parameter of the qualifier is$(0, 0.5)$.
  By applying Eq. (6.1) we get

$$w_1 = Q\left(\frac{1}{4}\right) - Q\left(\frac{0}{4}\right) = Q(0.25) = 0.5,$$

$$w_2 = Q\left(\frac{2}{4}\right) - Q\left(\frac{1}{4}\right) = Q(0.5) - Q(0.25) = \frac{0.5-0}{0.5-0} - 0.5 = 0.5,$$

$$w_3 = Q\left(\frac{3}{4}\right) - Q\left(\frac{2}{4}\right) = 1 - 1 = 0,$$

and $w_4 = Q(1) - Q\left(\frac{3}{4}\right) = 1 - Q(0.7) = 1 - 1 = 0.$

Hence the weighting vector is w=[0.5, 0.5, 0, 0]

Now we can aggregate the parameters using the *LOWA* operator with each weighting vector. The first two sub parameters of *MIS* have to be aggregated. For that, the weighting vector for each quantifier mentioned above needs to be generated for two parameters. Following a similar procedure to the one mentioned above, we get the weighting vectors for each quantifier as follows:

- Mean → [0.5,0.5]
- Most → [0.4,0.6]
- At Least half → [1,0]

Aggregating sub parameters of *MIS*

The performance values of the two sub parameters, *HA* and *NMD*, are combined using the *LOWA* operator for each IAPS data set. This value represents the performance values of the main parameter *MIS*. Using the three quantifiers mentioned above, three sets of performance values have been obtained for *MIS*.

For example, *Very High* (*H*), *Low* (*L*) labels are the performance values of the species *Alstonia macrophylla* parameters with respect to parameters *NMD* and *HA*. To aggregate these two values using the *LOWA* operator with the "Most" quantifier guide is as follows:

The performance values are already in descending order VH, L.
Since there are only two labels, only one pair is to be aggregated. The indexes of the labels VH and L are 5 and 2 respectively.
By applying *LOWA* for the pair (VH, L) with weighting vector [0,1] ("Most" quantifier) we get

$$k_2=\min\left\{6, 2+r\left(\frac{0.4}{1}\times 3\right)\right\} = M\ (j=5\text{and } i=2),$$

i.e. Medium is the performance value of *MIS* of the species *Alstonia macrophylla*.
If the quantifier is "At least half" with weight vector [1,0] we get

$$k_2=\min\left\{6, 2+r\left(\frac{1}{1}\times 3\right)\right\} = VH\ (j=5\text{and } i=2),$$

i.e. "Very High" as the performance value.
Likewise each plant species performance value w.r.t. *MIS* has been obtained.
Let us see the procedure to aggregate the main parameters.

Aggregating of main parameters One may observe that there are four main parameters to be aggregated. Four parameters DIS, VRS, SGL, and MIS are combined using the *LOWA* operator by separately applying the three weight vectors obtained from the quantifiers.

For example, let us consider the species *Mimosa invisa* having performance values "*Very Low*"(VL),"*High*"(H), "*Medium*"(M), and "*Medium*"(M) for the parameters DIS, VRS, SGL, MIS respectively.

First, the four performance values need to be placed in descending order as H, M, M, VL and the indexes of these labels are 4, 3, 3, and 1.

If the quantifier is "At least half" with weight vector [0.5, 0.5, 0, 0] then by applying the *LOWA* operator for the first pair of the ordered values from the right, i.e. (M, VL), we get

k_2=min$\{6, 1+r(0\times 2)\}$ =1= VL ($j=3$and $i=1$) and $w_2=0$.

Then combine the value of $k_2=VL$ with the second label of the ordered values, i.e. M as
k_3=min$\{6, 1+r(1\times 2)\}$ = 3= M ($j=3$and $i=1$) and $w_3=1$.

Now combine the value of $k_3=M$ with the first label of the ordered values, i.e. H as
k_4=min$\{6, 3+r(0.5\times 1)\}$ = 4= H ($j=4$and $i=3$) and $w_4=0.5$.

Hence the invasion risk of the species *Mimosa invisa* described by the four parameters is "*High*" if the quantifier is "At least half."

Now assume the quantifier is "Most" with weight vector $[0, 0.4, 0.5, 0.1]$. Then the aggregation is as follows:

By taking the first pair of the ordered values from the right, i.e. (M, VL), we get

k_2=min$\left\{6, 1+r\left(\frac{0.5}{0.6}\times 2\right)\right\}$ =3= M ($j=3$and $i=1$) and $w_2=\frac{0.5}{0.6}$.

Then combine the value of $k_2=M$ with the second label of the ordered values, i.e. M as
k_3=min$\{6, 3+r(0.4\times 0)\}$ = 3= M ($j=3$and $i=1$) and $w_3=0.4$.

Now combine the value of $k_3=M$ with the first label of the ordered values, i.e. H as
k_4=min$\{6, 3+r(0\times 1)\}$ = 3 = M ($j=4$and $i=3$) and $w_4=0.5$.

Table 10.2

Test Results of Model I – *LOWA*. (Reprinted from H.O.W., Peiris, S., Chakraverty, S.S.N., Perera, and S.M.W., Ranwala. "Novel fuzzy linguistic based mathematical model to assess risk of invasive alien plant species," *Applied Soft Computing* 59, (2017): 326-339, with permission from Elsevier.)

Invasive Species	Model I			NRA Score
	Mean	Most	At least half	
Alternanthera philoxeroides	*M*	*M*	*VH*	*H*
Clidemia hirta	*H*	*M*	*H*	*H*
Miconia calvescens	*VH*	*H*	*EH*	*H*
Alstonia macrophylla	*H*	*M*	*VH*	*M*
Annona glabra	*L*	*L*	*M*	*M*
Clusia rosea	*M*	*L*	*M*	*M*
Dillenia suffructicosa	*L*	*VL*	*H*	*M*
Ageratina riparia	*H*	*M*	*H*	*M*
Mimosa invisa	*H*	*M*	*H*	*H*
Myroxylon balsamum	*M*	*L*	*H*	*M*
Tithonia diversiflora	*M*	*L*	*H*	*M*
Mikania micrantha	*H*	*H*	*H*	*H*
Prosopis juliflora	*H*	*M*	*H*	*H*
Ulex europaeus	*M*	*L*	*M*	*M*
Mimosa pigra	*H*	*M*	*H*	*H*
Chromolaena odorata	*H*	*M*	*H*	*H*
Parthenium hysterophorus	*H*	*M*	*H*	*H*
Lantana camara	*M*	*M*	*M*	*M*
Imperata cylindrical	*H*	*H*	*VH*	*VH*
Opuntia stricta	*H*	*M*	*H*	*H*
Colubrina asiatica	*M*	*M*	*VH*	*H*
Pennisetum polystachion	*H*	*M*	*H*	*M*
Sphagneticola trilobata	*M*	*L*	*H*	*M*
Zizigium marutinum	*L*	*VL*	*M*	*L*
Eichhornia crassipes	*H*	*M*	*VH*	*H*
Pistia stratiotes	*H*	*M*	*VH*	*H*
Leucaena leucocephala	*M*	*L*	*H*	*M*

Thus the invasion risk of species *Mimosa invisa* is "Medium" if the quantifier is "*Most*."

Likewise, for each species in the data set, the performance values of four parameters have been combined by applying the three quantifiers separately. These results are presented in Table 10.2. The model has been validated using the same set of known invasive and non-invasive species that has been used in Chapter 9 (see Table 10.3).

Table 10.3
Validation results – Model I – *LOWA*. (Reprinted from H.O.W., Peiris, S., Chakraverty, S.S.N., Perera, and S.M.W., Ranwala. "Novel fuzzy linguistic based mathematical model to assess risk of invasive alien plant species," *Applied Soft Computing* 59, (2017): 326-339, with permission from Elsevier.)

Category of Species	Species	Model I			NRA
		Mean	Most	At least half	
Invasive	*Austroeupator iuminulifolium*	*M*	*M*	*H*	*M*
	Panicum maximum	*M*	*L*	*VH*	*H*
	Cuscuta campestris	*M*	*M*	*VH*	*H*
	Pueraria montana	*M*	*M*	*VH*	*H*
	Acacia mearnsii	*H*	*M*	*H*	*M*
	Myrica faya	*M*	*L*	*H*	*M*
Non-Invasive	*Cassia fistula*	*M*	*L*	*H*	*M*
	Cissus rotundifolia	*L*	*L*	*M*	*L*
	Hedychium gardnerianum	*L*	*VL*	*M*	*L*
	Magnefera indica	*L*	*L*	*H*	*M*

10.2.4 DISCUSSION – MODEL I

Test results in Table 10.2 reveal that the compatibility of model output with NRA risk levels may vary on the selected quantifier. The outputs of 18 species out of 27 species are compatible with NRA levels if the model executes with the "Mean" quantifier.

In contrast, the validation results in Table 10.3, risk levels of non-invasive species, are mostly compatible with NRA levels if the quantifier is "Most". However, in the invasive category, only the risk levels of species such as *Austroeupatorium inulifolium*, *Acacia mearnsii*, and *Myrica faya* are compatible with different quantifiers.

Below we present the model developed via *LWA* operator with weighted parameters.

10.2.5 MODEL II – *LWA*

In the previous model, the parameters are considered as equally important towards assessing the invasion risk of IAPS. In this task we assume that the main/sub parameters are not equally important towards invasiveness; i.e. each parameter should be accompanied with a weight. Therefore the model structure which is depicted in Figure 6.4. has been followed to evaluate invasion risk by adopting importance weights of the parameters. The first two steps and the step evaluation of performance values of the parameters may proceed in a similar way as the previous model. Now let us discuss the remaining steps of the model one by one starting with the step relating to evaluation of importance weights of the parameters below:

Table 10.4

Importance weights of main/sub parameters. (Reprinted from H.O.W., Peiris, S., Chakraverty, S.S.N., Perera, and S.M.W., Ranwala. "Novel fuzzy linguistic based mathematical model to assess risk of invasive alien plant species," *Applied Soft Computing* 59, (2017): 326-339, with permission from Elsevier.)

Importance Weight of Main parameter	Main parameter	Sub parameter	Weights of Sub parameter
Very High	*DIS*	-	-
Very High	*VRS*	-	-
Medium	*SGL*	-	-
High	*MIS*	*NMD*	*Medium*
		HA	*High*

Evaluation of Importance Weights The importance weights of the main and sub parameters are gathered from the group of three plant science experts. These opinions have been gathered as a group decision based on the linguistic scale as in Figure 4.4.
Table 10.4 presents the importance weights obtained for each main and sub parameter.

Combining importance weights and the performance values of the parameters In this step, the importance weight of the parameter needs to be aggregated with the corresponding performance values. Therefore, the classical MIN operator as stated in Subsection 6.3.2. has been used. For example, consider the species *Mimosa invisa* with performance values "*VeryLow*"(VL), "*High*"(H), "*Medium*"(M), and "*Medium*"(M) for the parameters DIS, VRS, SGL, MIS respectively.
If one needs to obtain the weighted performance value of the parameter DIS, combine the performance value "*Very Low*" with its corresponding weight "Very High" as
$LC_1^{\rightarrow}\,(Very\ High, Very\ Low) = MIN\,(Very\ High, Very\ Low)$="*Very Low*"$(VL)$.

Likewise the performance value of each main/sub parameter has to be combined with the corresponding importance weight.

Table 10.5 shows the values of weighted parameters (main) corresponding to the species *Mimosa invisa*

Aggregation of weighted parameters Here the weighted parameters need to be aggregated using the LWA operator with appropriate linguistic quantifier. First the weighted sub parameters of the main parameter MIS have to be aggregated; thereafter, the main four weighted parameters are aggregated.

Table 10.5
Values of weighted main parameters of *Mimosa invisa*

Main parameter	Performance value	Importance weight	Weighted value
DIS	*Very Low*	*Very High*	*Very Low*
VRS	*High*	*Very High*	*High*
SGL	*Medium*	*Medium*	*Medium*
MIS	*Medium*	*High*	*Medium*

In order to find the appropriate quantifier, one may go to the final step, i.e. validation. We already have the NRA risk levels of IAPS relating to the parameters considered in this model. For example, one can proceed with the "Mean" quantifier and check the compatibility with NRA levels. If the output of some/more species is not at a satisfactory level, then the quantifier may change into either "Most" or "At least half."

Now let us elaborate upon this task using an example:

Consider the species *Mimosa invisa* with wieghted performance values "*Very Low*" (VL), "*High*"(H), "*Medium*"(M), "*Medium*"(M) and "*High*"(H) for the parameters DIS, VRS, SGL, NMD, HA respectively.
First consider the pair (M,H) which represents the weighted performance values of the sub parameters NMD and HA respectively. If the quantifier is "Most" with weight vector [0.4,0.6] then the two sub parameters are aggregated using LWA as follows:

Here the indexes of the labels M and H are 3,4 respectively. $k_2 = \min\{6, 3 + r(0.4 \times 1)\} = M$ (Here $j = 4$ and $i = 3$).

In order to obtain the value $Medium$ as for the weighted performance value of the main parameter MIS, that label M needs to be combined with the corresponding weight H (see Table 10.4) using classical MIN as:

$LC_1^{\rightarrow}(M, High) = MIN(M, H) = M.$

Now one may see that the weighted performance value of the parameter MIS is "Medium."

Let us move to the next part, i.e. aggregation of the weighted main parameters DIS, VRS, SGL, MIS with performance values (VL, H, M, M).

First prepare the labels in descending order as:

(H, M, M, VL).
Now take the first pair from the right, i.e. (M, VL).

If the quantifier is "Most" with weight vector $[0, 0.4, 0.5, 0.1]$

$$k_2 = \min\left\{6, 1 + r\left(\frac{0.5}{0.6} \times 2\right)\right\} = M \text{ (Here } j = 3 \text{ and } i = 1).$$

Now take the second pair, i.e. (M, M),

$$k_3 = \min\{6, 3 + r(0.4 \times 0)\} = M \text{ (Here } j = 3 \text{ and } i = 3).$$

For the last pair (H, M)

$$k_4 = \min\left\{6, 3 + r\left(\frac{0}{1} \times 1\right)\right\} = M \text{ (Here } j = 4 \text{ and } i = 3).$$

Therefore, the aggregated risk level of the species *Mimosa invisa* is "Medium" under the quantifier "Most."

Likewise, with different quantifiers we have evaluated the risk levels of IAPS and checked the compatibility with NRA levels. In that task, we have noticed plant performances may vary according to the chosen quantifier. In that sense, this process has been optimized and has identified several plant categories and their suitable quantifier as follows:

Category I. Plant whose $DIS \geq Neg(VRP)$.
Here "Neg" denotes the negation operator and at least one linguistic value of DIS/VRP should be greater than or equal to the label "Medium."
The weighted parameters of the plant species that falls into this category are aggregated using *LWA* with "Mean" quantifier guide otherwise one proceeds with "At least half" as the quantifier guide.

Category II. Plant whose $DIS = High \; \& \; VRP = Medium \; or \; Low \; \& \; SGL = Medium$.

The weighted parameters of the plant species that falls into this category are aggregated using *LWA* with "At least half" quantifier guide.

Category III. Plant whose linguistic value of $DIS \; \& \; VRP$ is less than or equal to the label "Low."

The weighted parameters of the plant species that falls into this category are then aggregated using *LWA* with "Most" quantifier guide.

It can be seen that this model executes under three different plant categories. The model has also been validated using the same set used in Model I – *LOWA*. The test and validation results are presented in Tables 10.6 and 10.7.

Table 10.6
Test Results of Model II – *LWA*. (Reprinted from H.O.W., Peiris, S., Chakraverty, S.S.N., Perera, and S.M.W., Ranwala. "Novel fuzzy linguistic based mathematical model to assess risk of invasive alien plant species," *Applied Soft Computing* 59, (2017): 326-339, with permission from Elsevier.)

Invasive Species	Model	NRA
Alternanthera philoxeroides	*High*	*High*
Clidemia hirta	*High*	*High*
Miconia calvescens	*High*	*High*
Alstonia macrophylla	*Medium*	*Medium*
Annona glabra	*Medium*	*Medium*
Clusia rosea	*Medium*	*Medium*
Dillenia suffructicosa	*Medium*	*Medium*
Ageratina riparia	*Medium*	*Medium*
Mimosa invisa	*High*	*High*
Myroxylon balsamum	*Medium*	*Medium*
Tithonia diversiflora	*Low*	*Medium*
Mikania micrantha	*High*	*High*
Prosopis juliflora	*High*	*High*
Ulex europaeus	*Medium*	*Medium*
Mimosa pigra	*High*	*High*
Chromolaena odorata	*High*	*High*
Parthenium hysterophorus	*High*	*High*
Lantana camara	*Medium*	*Medium*
Imperata cylindrical	*High*	*VeryHigh*
Opuntia stricta	*High*	*High*
Colubrina asiatica	*Medium*	*Medium*
Pennisetum polystachion	*Medium*	*Medium*
Sphagneticola trilobata	*Medium*	*Medium*
Zizigium marutinum	*Low*	*Low*
Eichhornia crassipes	*High*	*High*
Pistia stratiotes	*High*	*High*
Leucaena leucocephala	*Medium*	*Medium*

10.2.6 DISCUSSION – MODEL II – *LWA*

Test results in Table 10.6 show that most of the species risk levels are compatible with NRA levels. Apart from that, in validation results, it can be seen that the risk levels of species in the non-invasive category are positioned in the "*Low*" risk level. This fact reveals the significance of this model compared to Model I – *LOWA*.

Below we discuss the model based on the operator *MLIOWA*.

10.2.7 MODEL III – *MLIOWA*

Here we have followed the model structure as depicted in Figure 6.5 to build up the model based on the operator *MLIOWA*.

Table 10.7
Validation results – Model II – *LWA*. (Reprinted from H.O.W., Peiris, S., Chakraverty, S.S.N., Perera, and S.M.W., Ranwala. "Novel fuzzy linguistic based mathematical model to assess risk of invasive alien plant species," *Applied Soft Computing* 59, (2017): 326-339, with permission from Elsevier.)

Category of Species	Species	Model II	NRA
Invasive	*Austroeupator iuminulifolium*	*Medium*	*Medium*
	Panicum maximum	*High*	*High*
	Cuscuta campestris	*High*	*High*
	Pueraria montana	*High*	*High*
	Acacia mearnsii	*Medium*	*Medium*
	Myrica faya	*Medium*	*Medium*
Non-Invasive	*Cassia fistula*	*Low*	*Medium*
	Cissus rotundifolia	*Low*	*Low*
	Hedychium gardnerianum	*Low*	*Low*
	Magnefera indica	*Low*	*Medium*

One may note that the first four steps, identifying parameters, defining the cardinality, evaluating the importance weights, and the performance values of the parameters need to be executed in a similar way to Model II – *LWA*.

Therefore, let us start with the step evaluation of sup_i.

Evaluate sup_i: As mentioned in Chapter 6, sup_i is calculated w.r.t. importance weights. The sup_i values for the importance weights of the main parameters have been calculated according to Definition (6.4). Here we have chosen α as 1. Using these sup_i values, the order inducing values u_i have also been evaluated. The values sup_i and u_i are presented in Table 10.8 (Peiris, Chakraverty, Perera and Ranwala, (2017) 326-339).

Evaluate w_i: Here we consider two linguistic quantifiers such as "Most" and "At least half" to obtain two different weight sets. If the quantifier is "At least half", the weight vector is calculated as follows:

$$w_1 = Q\left(\frac{2}{4}\right) / \Sigma_{j=1}^{n} Q\left(\frac{u_{\sigma(j)}}{n}\right) = \frac{0.4}{3.05} \text{ and}$$

$$w_2 = Q\left(\frac{2.5}{4}\right) / \Sigma_{j=1}^{n} Q\left(\frac{u_{\sigma(j)}}{n}\right) = \frac{0.65}{3.05}.$$

$$w_3 = w_4 = Q\left(\frac{3.5}{4}\right) / \Sigma_{j=1}^{n} Q\left(\frac{u_{\sigma(j)}}{n}\right) = \frac{1}{3.05}.$$

Table 10.8

Order inducing values of importance weights of main risk factors. (Reprinted from H.O.W., Peiris, S., Chakraverty, S.S.N., Perera, and S.M.W., Ranwala. "Novel fuzzy linguistic based mathematical model to assess risk of invasive alien plant species," *Applied Soft Computing* 59, (2017): 326-339, with permission from Elsevier.)

Main Risk Factor	Importance Weights	sup_i	u_i
DIS	*Very High*	2	3.5
VRS	*Very High*	2	3.5
SGL	*Medium*	1	2
MIS	*High*	1	2.5

Thus, the weight vector is $[\frac{0.4}{3.05}, \frac{0.65}{3.05}, \frac{1}{3.05}, \frac{1}{3.05}]$.

Let us now discuss the step corresponding to the parameter aggregation.

Aggregation of parameters To discuss the aggregation procedure, consider the invasive species *Miconia calvescens*'s performance values with importance weights $((VH,VH),(VH,VL),(M,H),(H,M))$ for the main risk factors *DIS*, *VRS*, *SGL*, *MIS* respectively. One may note that the first element in each pair indicates the importance weight. Using the weight vector obtained in the previous step, the four parameters are aggregated as follows:

First rearrange the pairs in ascending order by importance weight as:

$((M,H),(H,M),(VH,VH),(VH,VL))$.

Then

$$\Phi^{I}_{Q_1}((M,H),(H,M),(VH,VH),(VH,VL))$$
$$= \text{round}\left(4 * \frac{0.4}{3.05} + 3 * \frac{0.65}{3.05} + 5 * \frac{1}{3.05} + 5 * \frac{1}{3.05}\right)$$
$$= \text{round}(3.97) = 4 = H = High, \text{ i.e.}$$

invasion risk of *Clidemia hirta* is "High".

In the final step, we can check the compatibility with NRA values. To eliminate any considerable changes, the quantifier may be changed accordingly.

The test results of the model are presented in Table 10.9. The model has been validated using the same set of species as in Model II – *LWA* and the results are presented in Table 10.10.

One may note that the order inducing values u_i for the importance weights of sub factors (*NMD* and *HA*) of main factor *MIS* have been evaluated using the "Most" linguistic quantifier and the performance values of *MIS* have been evaluated using the weighted *MLIOWA* operator.

Table 10.9

Test Results of Model III – *MLIOWA*. (Reprinted from H.O.W., Peiris, S., Chakraverty, S.S.N., Perera, and S.M.W., Ranwala. "Novel fuzzy linguistic based mathematical model to assess risk of invasive alien plant species," *Applied Soft Computing* 59, (2017): 326-339, with permission from Elsevier.)

Invasive Species	Model III		NRA Score
	Most	At least half	
Alternanthera philoxeroides	*High*	*High*	*High*
Clidemia hirta	*Medium*	*Medium*	*High*
Miconia calvescens	*High*	*High*	*High*
Alstonia macrophylla	*Medium*	*Medium*	*Medium*
Annona glabra	*Low*	*Medium*	*Medium*
Clusia rosea	*Low*	*Medium*	*Medium*
Dillenia suffructicosa	*Medium*	*Medium*	*Medium*
Ageratina riparia	*High*	*High*	*Medium*
Mimosa invisa	*Medium*	*Medium*	*High*
Myroxylon balsamum	*Medium*	*Medium*	*Medium*
Tithonia diversiflora	*Medium*	*Medium*	*Medium*
Mikania micrantha	*High*	*High*	*High*
Prosopis juliflora	*Medium*	*High*	*High*
Ulex europaeus	*Low*	*Medium*	*Medium*
Mimosa pigra	*Medium*	*Medium*	*High*
Chromolaena odorata	*Medium*	*High*	*High*
Parthenium hysterophorus	*High*	*High*	*High*
Lantana camara	*Medium*	*Medium*	*Medium*
Imperata cylindrical	*High*	*High*	*Very High*
Opuntia stricta	*Medium*	*Medium*	*High*
Colubrina asiatica	*Medium*	*Medium*	*Medium*
Pennisetum polystachion	*High*	*High*	*Medium*
Sphagneticola trilobata	*Medium*	*Medium*	*Medium*
Zizigium marutinum	*Low*	*Low*	*Low*
Eichhornia crassipes	*High*	*High*	*High*
Pistia stratiotes	*High*	*High*	*High*
Leucaena leucocephala	*Medium*	*Medium*	*Medium*

10.2.8 DISCUSSION – MODEL III

According to Table 10.9, it can be clearly seen that the model produces significant results if the quantifier is "At least half." However the risk levels of some of the species in the non-invasive category outreached the expected levels.
Unlike Model II – *LWA*, there is no possibility to classify the plant groups in a logical sequence.

Therefore, comparing the results of Model II and III, it can be concluded that Model II with the LWA operator is the appropriate one to evaluate the risk of IAPS if these species' risk is dominated by the above mentioned parameters.

Table 10.10

Validation results – Model III – *MLIOWA*. (Reprinted from H.O.W., Peiris, S., Chakraverty, S.S.N., Perera, and S.M.W., Ranwala. "Novel fuzzy linguistic based mathematical model to assess risk of invasive alien plant species," *Applied Soft Computing* 59, (2017): 326-339, with permission from Elsevier.)

Category of Species	Species	Model III		NRA
		Most	At least half	
Invasive	*Austroeupator iuminulifolium*	*Medium*	*Medium*	*Medium*
	Panicum maximum	*Medium*	*Medium*	*High*
	Cuscuta campestris	*Medium*	*High*	*High*
	Pueraria montana	*High*	*High*	*High*
	Acacia mearnsii	*High*	*Medium*	*Medium*
	Myrica faya	*Medium*	*Medium*	*Medium*
Non-Invasive	*Cassia fistula*	*Medium*	*Medium*	*Medium*
	Cissus rotundifolia	*Medium*	*Low*	*Low*
	Hedychium gardnerianum	*Low*	*Low*	*Low*
	Magnefera indica	*Low*	*Medium*	*Medium*

10.3 MODELS – FUZZY 2-TUPLE APPROACH

In the previous section, we modeled the invasion risk by incorporating fuzzy majority guided averaging operators. Since the rounding off operator is accompanied by these operators, the actual result may not emerge. Due to that fact, in this task we have followed the model structure given in Figure 6.6 to develop invasion risk models based on the 2-tuple approach. Here we develop two models considering the equal/unequal importance of the parameters.

Below we discuss development of the risk model accompanied by equally important parameters.

10.3.1 MODEL I – 2-TUPLE (NON-WEIGHTED)

Here we model the invasion risk considering the same set of parameters as in the previous section. We have considered these parameters as equally important towards invasiveness, i.e. parameters are non-weighted. Below, we discuss the development steps of this model in detail.

One may observe that the first two steps are similar to the previous models. Therefore, let us begin with the third step as follows:

Evaluation of performance values: In this case, what we need is to transform the performance values of the parameters already obtained from IAPS into 2-tuples.

For example, the species *Alstonia macrophylla*'s performance values of the parameters DIS, VRS, SGL, HA, NMD are VL, H, L, M, and VH respectively. These values have been gathered directly from the linguistic scale defined in Figure 4.4. Therefore, the symbolic translation value α is 0. Hence the 2-tuple representation of the performance values of the above mentioned species is:

$$((VL,0),(H,0),(L,0),(M,0),(VH,0)).$$

In the same way, the performance values of the parameters w.r.t. each species in the data set have been converted into 2-tuples.

Combining of performance values: This step relates to the aggregation of the parameters.

The parameters can be aggregated directly through Eq. (6.4) since we have assumed that the parameters are equally important. Apart from that, one need not evaluate the performance value of the main parameter *MIS* separately as its two sub parameters can be combined with main parameters directly.

For instance, consider the species *Alstonia macrophylla*'s 2-tuple performance values $((VL,0),(H,0),(L,0),(M,0),(VH,0))$ of the parameters *DIS*, *VRS*, *SGL*, *HA*, *NMD* respectively. The indexes of the labels of corresponding performance values may be presented as:

$$((1,0),(4,0),(2,0),(3,0),(5,0))$$

Using Eq. (6.4) we get

$$\frac{(1+4+2+3+5)}{5}=3$$

As to Eq. (6.3), β is 3. Since the difference between the rounded off value of β with itself is zero, then the value of α is 0. And the corresponding label s_i is "Medium." Therefore the 2-tuple for the aggregated risk of the species *Alstonia macrophylla* is $[M,0]$.

In the same way, the invasion risk of each species in the data set has been evaluated. Further, the model has been validated using the same set which has been used to validate the previous models. The results are presented in Tables 10.11 and 10.12.

10.3.2 DISCUSSION – MODEL I – 2-TUPLE (NON-WEIGHTED)

According to Table 10.11, it can be seen that the risk of species such as *Alstonia macrophylla*, *Lantana camara* produced from the model are exactly the same as NRA. Apart from that, for some species, the NRA and linguistic label s_i in the 2-tuple is the same but different in symbolic translation (α).

For example, consider the 2-tuples of species *Dillenia suffructicosa* and *Ageratina riparia* as depicted in Figure 10.1.

Table 10.11
Test results of Model I – 2-tuple – Non-weighted

Invasive Species	NRA	Model Output
Alternanthera philoxeroides	*High*	[*H*, -0.4]
Clidemia hirta	*High*	[*M*, 0]
Miconia calvescens	*High*	[*H*, -0.4]
Alstonia macrophylla	*Medium*	[*M*, 0]
Annona glabra	*Medium*	[*M*, -0.4]
Clusia rosea	*Medium*	[*L*, 0.4]
Dillenia suffructicosa	*Medium*	[*M*, -0.4]
Ageratina riparia	*Medium*	[*M*, 0.4]
Mimosa invisa	*High*	[*M*, 0]
Myroxylon balsamum	*Medium*	[*M*, -0.4]
Tithonia diversiflora	*Medium*	[*M*, -0.2]
Mikania micrantha	*High*	[*H*, -0.2]
Prosopis juliflora	*High*	[*M*, 0.4]
Ulex europaeus	*Medium*	[*L*, 0.4]
Mimosa pigra	*High*	[*M*, 0]
Chromolaena odorata	*High*	[*M*, 0.4]
Parthenium hysterophorus	*High*	[*M*, 0.4]
Lantana camara	*Medium*	[*M*, 0]
Imperata cylindrical	*Very High*	[*H*, 0]
Opuntia stricta	*High*	[*M*, 0.2]
Colubrina asiatica	*Medium*	[*M*, 0.4]
Pennisetum polystachion	*Medium*	[*M*, 0.4]
Sphagneticola trilobata	*Medium*	[*M*, -0.2]
Zizphus mauritiana	*Low*	[*L*, 0.2]
Eichhornia crassipes	*High*	[*M*, 0.4]
Pistia stratiotes	*High*	[*M*, 0.4]
Leucaena leucocephala	*Medium*	[*M*, -0.2]

Table 10.12
Validation results – Model I – 2-tuple – Non-weighted

Category of Species	Species	NRA	Model Output
Invasive	*Austroeupator iuminulifolium*	*Medium*	[*M*, 0]
	Panicum maximum	*High*	[*M*, 0.2]
	Cuscuta campestris	*High*	[*M*, 0.4]
	Pueraria montana	*High*	[*H*, -0.2]
	Acacia mearnsii	*Medium*	[*M*, 0]
	Myrica faya	*Medium*	[*M*, -0.4]
Non-Invasive	*Cassia fistula*	*Medium*	[*M*, -0.2]
	Cissus rotundifolia	*Low*	[*L*, 0.4]
	Hedychium gardnerianum	*Low*	[*L*, 0.2]
	Magnefera indica	*Medium*	[*M*, -0.4]

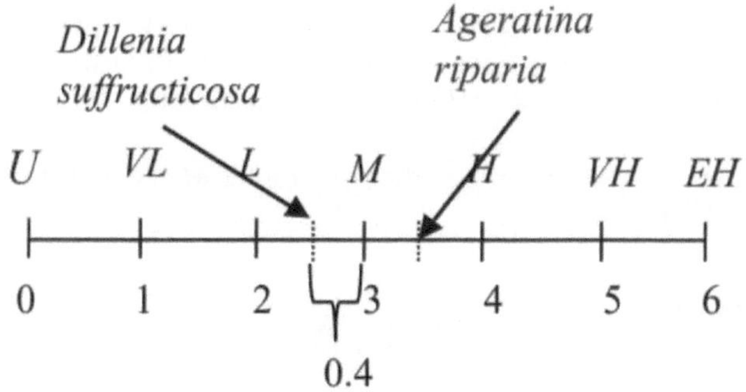

Figure 10.1 Comparison of Model I – 2-tuple (Non-weighted) output with NRA risk levels.

Figure 10.1. clearly interprets that *Ageratina riparia* is more aggressive than *Dillenia suffructicosa* as its risk stands 0.4 beyond the label "Medium" whereas *Dillenia suffructicosa* stands 0.4 behind the same label. Similarly, the risks of the species, *Alternanthera philoxeroides* and *Mikania micrantha* can be compared.

Let us consider the species in Table 10.12, where the risk labels in the 2-tuples are different from NRA label. The species such as *Panicum maximum* and *Cuscuta campestris* show "High" risk in NRA but obtained "Medium" from the model. But the value of symbolic translation is between the labels "Medium" and "High."

Let us consider the non-invasive species *Cassia fistula* and *Magnfera indica* which show "Medium" risk in NRA as we expect "Low" or below risk levels in reality. From this model, it can be clearly seen that the values of symbolic translation of *Cassia fistula* and *Magnefera indica* stay behind the "Medium" risk level 0.2 and 0.4 respectively.

In the following subsection we present the model based on the 2-tuple approach with weighted parameters.

10.3.3 MODEL II – 2-TUPLE (WEIGHTED)

Here we model the invasion risk based on the structure depicted in Figure 6.6 by incorporating weighted main/sub parameters as presented in Table 10.4. The weights of the parameters are evaluated using linguistic quantifiers, "Most", "Mean", and "At least half." The aggregation process has been executed as follows:

- Initially, the sub parameters with weights have been combined to evaluate the performance value of *MIS*.

For example, consider the species *Clusia rosea*'s performances values (*M*,*L*) with respect to sub parameters *NMD*, *HA* respectively. If the quantifier is "Most" with weighting vector [0.4,0.6], using Eq. (6.5) the performance value of the main parameter *MIS* is evaluated as follows:

First rearrange the labels in ascending order as (*L*,*M*). Note that the indexes of the labels of *L* and *M* are 2, 3 respectively. Using Eq. (6.5) we get

$$\Delta\left(\frac{\sum_{i=1}^{n}\beta_i w_i}{\sum_{i=1}^{n} w_i}\right) = \frac{(2\times 0.4+3\times 0.6)}{1} = 2.6.$$

Here $\beta = 2.6$ and its rounded off value is 3, i.e. $i = 3$. Hence $s_i = Medium$ and $\alpha = \beta - \alpha = 2.6 - 3 = -0.4$. The performance value of the species *Clusia rosea* w.r.t. *MIS* is $[M, -0.4]$.

Likewise, the performance value of each invasive species in the data set w.r.t. *MIS* has been evaluated using linguistic quantifiers "Most" and "At least half."

- Finally, the performance values of the main four parameters have been combined with weights obtained from quantifiers "Most" and "At least half."
  For example, consider the performance values of species *Opuntia stricta* $((H,0),(H,0),(L,0),(L,-0.2))$ w.r.t. parameters *DIS*, *VRS*, *SGL*, *MIS* respectively. If the quantifier is "Most," the combination of corresponding weight vector and performance values of the parameters is carried out as follows:

$$\Delta\left(\frac{\sum_{i=1}^{n}\beta_i w_i}{\sum_{i=1}^{n} w_i}\right) = \frac{(4\times 0+4\times 0.4+2\times 0.5+0.1\times 1.8)}{1} = 3.2.$$

According to the value obtained above, $s_i = Medium$ and $\alpha = 0.2$. The risk of *Opuntia stricta* is $[M, 0.2]$.

In a similar way, the risk of each species in the data set has been obtained. Further, model has been validated using the same set of species used in the previous model. The results are presented in Tables 10.13 and 10.14.

10.3.4 DISCUSSION – MODEL II – 2-TUPLE (WEIGHTED)

According to the results given in Table 10.13, the species' risk obtained from the model with the "At least half" quantifier is compatible with corresponding NRA levels except the species *Miconia calvescens*, *Alstonia macrophylla*, *Annona glabra*, *Ageratina riparia*, *Pennisetum polystachion*, and *Zizigium marutinum*.

Table 10.13

Test Results of Model II – 2-tuple (Weighted)

Invasive Species	Model Output		NRA Score
	Most	At least half	
Alternanthera philoxeroides	[*H*, -0.22]	[*VH*, -0.5]	*High*
Clidemia hirta	[*M*, 0.4]	[*H*, 0]	*High*
Miconia calvescens	[*H*, -0.5]	[*EH*, 0]	*High*
Alstonia macrophylla	[*H*, 0.02]	[*VH*, 0]	*Medium*
Annona glabra	[*M*, -0.32]	[*M*, 0]	*Medium*
Clusia rosea	[*M*, -0.16]	[*M*, 0]	*Medium*
Dillenia suffructicosa	[*L*, 0.14]	[*H*, -0.5]	*Medium*
Ageratina riparia	[*H*, 0]	[*H*, 0]	*Medium*
Mimosa invisa	[*H*, -0.16]	[*H*, 0]	*High*
Myroxylon balsamum	[*M*, -0.06]	[*H*, -0.5]	*Medium*
Tithonia diversiflora	[*M*, -0.5]	[*H*, -0.5]	*Medium*
Mikania micrantha	[*H*, 0]	[*H*, 0]	*High*
Prosopis juliflora	[*H*, -0.16]	[*H*, 0]	*High*
Ulex europaeus	[*M*, -0.16]	[*M*, 0]	*Medium*
Mimosa pigra	[*M*, 0.4]	[*H*, 0]	*High*
Chromolaena odorata	[*H*, -0.16]	[*H*, 0]	*High*
Parthenium hysterophorus	[*M*, 0.4]	[*H*, 0]	*High*
Lantana camara	[*M*, 0]	[*M*, 0]	*Medium*
Imperata cylindrical	[*H*, 0.1]	[*VH*, -0.5]	*Very High*
Opuntia stricta	[*M*, 0.2]	[*H*, 0]	*High*
Colubrina asiatica	[*M*, -0.34]	[*H*, -0.5]	*Medium*
Pennisetum polystachion	[*M*, 0.2]	[*H*, 0]	*Medium*
Sphagneticola trilobata	[*M*, -0.34]	[*H*, -0.5]	*Medium*
Zizigium marutinum	[*L*, 0.4]	[*M*, 0]	*Low*
Eichhornia crassipes	[*M*, 0.3]	[*VH*, -0.5]	*High*
Pistia stratiotes	[*H*, -0.16]	[*VH*, -0.5]	*High*
Leucaena leucocephala	[*M*, -0.06]	[*H*, -0.5]	*Medium*

Table 10.14

Validation results – Model II – 2-tuple (Weighted)

Category of Species	Species	Model Output		NRA
		Most	At least half	
Invasive	*Austroeupator iuminulifolium*	[*M*, 0.3]	[*H*, -0.5]	*Medium*
	Panicum maximum	[*M*, 0.1]	[*VH*, -0.5]	*High*
	Cuscuta campestris	[*H*, -0.22]	[*VH*, -0.5]	*High*
	Pueraria montana	[*H*, 0.1]	[*VH*, -0.5]	*High*
	Acacia mearnsii	[*H*, -0.4]	[*H*, 0]	*Medium*
	Myrica faya	[*M*, -0.34]	[*H*, -0.5]	*Medium*
Non-Invasive	*Cassia fistula*	[*L*, 0.1]	[*M*, -0.5]	*Medium*
	Cissus rotundifolia	[*M*, -0.4]	[*M*, 0]	*Low*
	Hedychium gardnerianum	[*M*, -0.06]	[*H*, -0.5]	*Low*
	Magnefera indica	[*M*, -0.34]	[*H*, -0.5]	*Medium*

In contrast, the performance of species in the non-invasive category are significant with the quantifier "Most."

Therefore, the model should execute with the "Most" quantifier if the terms of both *VRS* and $DIS \leq Low$; otherwise execution may proceed with the "At least half" quantifier.

10.4 MODELS – APPROXIMATION APPROACH

In this section, we discuss the models which have been developed based on the approximation approach mentioned in Section 6.6. Here we have considered two different linguistic scales for performance values and importance weights of the parameters.

Below we discuss the steps of the first model in detail.

10.4.1 MODEL I – APPROXIMATION APPROACH

Identify the parameters: Here we have added up four more parameters other than in previous models. These four parameters are Physical Defensive Structures (*PDS*), Formation of Dense Thickets (*FDS*), Alleopathy Property (*AP*), and Existence of Invasive Races (*IR*).

Unlike the other parameter values *PDS*, *FDS*, *AP*, and *IR* are in the form of bi-valued answers such as "Yes" or "No." We have chosen the linguistic terms set as depicted in Figure 4.4 for evaluating the performance values. One may observe that these two terms do not match with the linguistic terms set in Figure 4.4. Therefore, these two values have been matched with the linguistic terms set. The term "No" has been matched with linguistic term "Unlikely." The term "Yes" has been matched with the linguistic term "High" instead of the terms "Very High" and "Extremely High." This decision has been made to minimize the effects that occur from overestimated overall output.
Now we have eight parameters such as *DIS*, *VRS*, *SGL*, *HA*, *NMD*, *PDS*, *FDS*, *AP*, and *IR*. By gathering experts' opinions these parameters have been classified into main and sub parameters as presented in Table 10.15.

Linguistic terms set for importance weights: Here we have used a linguistic scale which is more supportive for decision makers to make their judgments. Figure 10.2 illustrates the chosen linguistic scale and Table 10.16 shows that the

Table 10.15

Parameters of Model I – Approximation Approach. (Adapted from H.O.W. Peiris, S.S.N. Perera, S. Chakraverty and S.M.W. Ranwala, "Development of a mathematical model to evaluate the rate of aggregate risk of invasive alien species: Fuzzy risk assessment approach," *International Journal of Bio Mathematics,* 11, no(4) (2018): 1-17.)

Main risk factor	Sub risk factor
Dispersal(*DIS*)	
Growth (*G*)	Vegetative reproductive strength (*VRS*)
	Formation of dense thickets (*FDS*)
	Physical defensive structures (*PDS*)
Seed germination requirement level (*SGL*)	
Alleopathy property (*AP*)	
Invasive races (*IR*)	
Man's influence on spreading (*MIS*)	Potential to be spread by human activities (*HA*)
	Role of natural and man made disturbances (*NMD*)

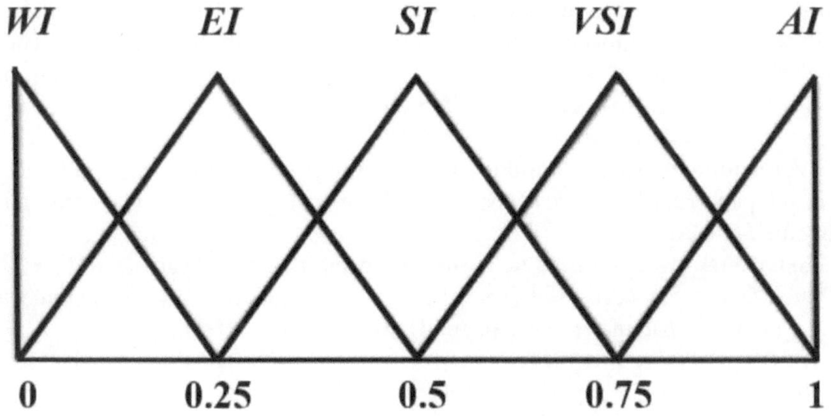

Figure 10.2 Membership functions of linguistic scale for importance weights. (Adapted from H.O.W. Peiris, S.S.N. Perera, S. Chakraverty and S.M.W. Ranwala, Development of a mathematical model to evaluate the rate of aggregate risk of invasive alien species: Fuzzy risk assessment approach," *International Journal of Bio Mathematics,* 11, no(4) (2018): 1-17.)

cardinality and semantics of the labels are different from the scale as depicted in Figure 4.4.

Collecting responses for importance weights from experts: The linguistic scale mentioned in the previous step has been used to gather the experts' opinions.

Table 10.16

Parameters of Model I – Approximation Approach. (Adapted from H.O.W. Peiris, S.S.N. Perera, S. Chakraverty and S.M.W. Ranwala, "Development of a mathematical model to evaluate the rate of aggregate risk of invasive alien species: Fuzzy risk assessment approach," *International Journal of Bio Mathematics,* 11, no(4) (2018): 1-17.)

Label	Linguistic term	Triangular fuzzy number
s_4	Absolutely more important (AI)	(0.75,1,1)
s_3	Very strongly more important (VSI)	(0.5,0.75,1)
s_2	Strongly more important (SI)	(0.25,0.5,0.75)
s_1	Equally important (EI)	(0,0.25,0.5)
s_0	Weakly more important (WI)	(0,0,0.25)
	Just equal	(1,1,1)

First, pairwise comparison of sub risk factors with respect to their main factor and pairwise comparisons among six main factors towards invasion risk have been obtained. A questionnaire form has been constructed as depicted in Figures 4.6 and 4.7 to obtain the decision makers' pairwise comparisons among the parameters. Three plant science experts' opinions have been provided for the pairwise comparisons using the linguistic scale as presented in Table 10.16.

Aggregation of experts' opinions The opinions that have been collected for each main/sub parameter from three experts have been aggregated to a single opinion by using the *LOWA* operator.

First, the responses for each main/sub parameter have been aggregated. Then the operator *LOWA* has been used to aggregate the comparison values of parameters w.r.t. a particular parameter (Herrera and Herrera-Viedma, 2000).

This process may be explained as follows:

For instance, consider the sub parameter vegetative reproduction strength (VRS) of the main factor Growth. The pairwise comparisons of VRS with respect to remaining sub factors of Growth obtained from three expert are as in Table 10.17.

One may observe that values of columns w.r.t. VRS, FDS, and PDS of Table 10.17 are aggregated by LOWA with "As many as possible" linguistic quantifier guide with weight vector $[\frac{2}{3},\frac{1}{3},0]$.

Now consider the labels of column FDS with respect to VRS as

$\{(\text{Expert } 1, VSI), (\text{Expert } 2, EI), (\text{Expert } 3, SI)\}$.

By preparing the labels in descending order we have

(VSI, SI, EI).

Table 10.17

Expert pairwise comparisons on *VRS* with respect to *FDS* and *PDS*. (Adapted from H.O.W. Peiris, S.S.N. Perera, S. Chakraverty and S.M.W. Ranwala, "Development of a mathematical model to evaluate the rate of aggregate risk of invasive alien species: Fuzzy risk assessment approach," *International Journal of Bio Mathematics*, 11, no(4) (2018): 1-17.)

		VRS	*FDS*	*PDS*
Expert 1	*VRS*	*JE*	*VSI*	*VSI*
Expert 2	*VRS*	*JE*	*EI*	*EI*
Expert 3	*VRS*	*JE*	*SI*	*EI*

Table 10.18

Aggregated pairwise comparisons of *VRS* with respect to *FDS* and *PDS*. (Adapted from H.O.W. Peiris, S.S.N. Perera, S. Chakraverty and S.M.W. Ranwala, "Development of a mathematical model to evaluate the rate of aggregate risk of invasive alien species: Fuzzy risk assessment approach," *International Journal of Bio Mathematics*, 11, no(4) (2018): 1-17.)

	VRS	*FDS*	*PDS*
VRS	-	*VSI*	*SI*

First consider the pair *SI* and *EI*. Applying LOWA
$k_2 = min\{4, 1 + r(1 \times 1)\} = SI$ (Here $j = 2$ and $i = 1$).
For the last pair *VSI* and *SI*
$k_3 = min\{4, 3 + r(\frac{2}{3} \times 1)\} = VSI$ (Here $j = 3$ and $i = 2$).

A similar procedure may be carried out for the remaining columns. Table 10.18 presents the aggregated values of each column.

To evaluate the final weight by aggregating the two labels (*VSI*,*SI*) as in Table 10.18 w.r.t. *VRS*, the LOWA with "As many as possible" quantifier guide with weighting vector $w = [1, 0]$ is used. Therefore the importance weight of *VRS* is *VSI*. Likewise, the importance weight for each main/sub parameter has been evaluated using different linguistic quantifiers such as "Most", "At least half", "As many as possible" and "Mean." The importance weights evaluated from the "As many as possible" quantifier are tabulated in Table 10.19.

Aggregating parameters by the merging method The importance weights and the corresponding performance values of the main/sub parameters have been merged and these weighted main/sub parameters have been aggregated by following the method mentioned in Subsection 6.6.1.

Table 10.19

Grade of importance weights of main/sub parameters. (Adapted from H.O.W. Peiris, S.S.N. Perera, S. Chakraverty and S.M.W. Ranwala, "Development of a mathematical model to evaluate the rate of aggregate risk of invasive alien species: Fuzzy risk assessment approach," *International Journal of Bio Mathematics,* 11, no(4) (2018): 1-17.)

Main Parameter	Grade of Importance Weight	Sub Parameter	Grade of Importance Weight
DIS	Very strongly more important		
G	Very strongly more important	*VRS*	Very strongly more important
		FDS	Very strongly more important
		PDS	Strongly more important
SGL	Strongly more important		
AP	Strongly more important		
IR	Very Strongly more important		
MIS	Equally important	*HA*	Very strongly more important
		NMD	Weakly more important

GW_{Mj} denotes the numerical value of the importance weight of the jth main parameter. Since the importance weights of main parameters are in the form of linguistic labels, the corresponding numerical value for each importance weight has been obtained by using the center of gravity method (Bai and Wang, 2006). These values are presented in Table 10.20.

The risks of 27 invasive species as used in previous models have been evaluated by following the latter part of the above mentioned method. Further, the model has been validated using the same set of known invasive and non-invasive species used in the previous models. Here the test and validation results are significant with the "As many as possible" quantifier and are presented in Tables 10.21 and 10.22.

10.4.2 DISCUSSION MODEL I – APPROXIMATION APPROACH

According to Table 10.21 it can be clearly seen that the aggregated risks of IAPS which take the NRA risk level *High* or above, *Medium*, and *Low* appear in the range of 0.2655 – 0.5722, 0.197 – 0.4487, 0.174 – 0.3267 respectively. The range 0.174 – 0.3267 is a subset of NRA risk level *Low* as its semantic ranges from 0.16 to 0.5.

On the other hand, the left end point of the model output range 0.197 – 0.448, i.e. 0.197, is 0.147 behind the expected left end point of the *Medium* risk level, i.e. 0.34. This fact has affected some species such as *Ulex europaeus*, *Annona glabra*, *Alternanthera philoxeroides*, *Pueraria montana*, where the output is slightly away from 0.34.

But the right end point of the output range is within the expected level. One may see that the left end point of the model output range 0.2655 – 0.5722 is the point 0.2345 behind the left end of the NRA risk level *High* but the right end point is within the specific range.

Table 10.20
Grades of importance weights of main parameters. (Adapted from H.O.W. Peiris, S.S.N. Perera, S. Chakraverty and S.M.W. Ranwala, "Development of a mathematical model to evaluate the rate of aggregate risk of invasive alien species: Fuzzy risk assessment approach," *International Journal of Bio Mathematics,* 11, no(4) (2018): 1-17.)

Main Parameters	GW_{Mj}
Dispersal	0.214286
Growth	0.214286
Seed germination requirement level	0.142857
Alleopathy property	0.142857
Invasive races	0.214286
Man's influence on spreading	0.071429

Due to that fact, the deviations have occurred for some species such as *Mimosa invisa*, *Prosopis juliflora*, *Opuntia stricta*, *Lantana camara*, and *Imperata cylindrical*. However, as per the validation results all the non-invasive species rates are *Low* and found within the expected risk level.

Next we discuss the second model which may be considered as the improved version of the first model.

10.5 MODEL II – APPROXIMATION APPROACH

This model has been developed by improving Model I, to minimize the deviations that occurred in that model. Here we introduce a new weight vector to weight the performance values of main/sub parameters.

The weighting mechanism may be explained as follows:

Mechanism to find W **and final output** Let us denote W as the weights given for main/sub risk factors.
These weights are in the form of linguistic labels obtained from the linguistic scale as depicted in Figure 4.4.

There is no unique way to find these weights; thus we randomly selected Ws which are taken from the scale as in Figure 4.4. These selected weights have been aggregated with performance values of main $((PM_j)_i)$ and sub $((PS_m)_j)$ parameters using the *LOWA* operator under a different linguistic quantifier. Afterwards, following the same steps as Model I, the risk of IAPS may be evaluated. The suitable W corresponding to each parameter has been chosen when the minimum deviation occurs among the model output and NRA risk levels.

Note that except for the weighted performance values of main/sub parameters other steps of Model I and II are the same. The finalized weights W are tabulated in

Table 10.21

Test results of Model I – Approximation Approach. (Adapted from H.O.W. Peiris, S.S.N. Perera, S. Chakraverty and S.M.W. Ranwala, "Development of a mathematical model to evaluate the rate of aggregate risk of invasive alien species: Fuzzy risk assessment approach," *International Journal of Bio Mathematics,* 11, no(4) (2018): 1-17.)

Invasive Species	Risk Level (NRA)	Aggregate Risk
Mikania micrantha	*High*	0.572154395
Chromolaena odorata	*High*	0.568820784
Mimosa pigra	*Very High*	0.555955808
Pennisetum polystachion	*High*	0.531504325
Leucaena leucocephala	*High*	0.52704379
Colubrina asiatica	*Medium*	0.448739556
Sphagneticola trilobata	*Medium*	0.422682288
Miconia calvescens	*High*	0.350024798
Mimosa invisa	*High*	0.337120182
Ageratina riparia	*Medium*	0.336364329
Clusia rosea	*Low*	0.326705113
Prosopis juliflora	*High*	0.325612544
Opuntia stricta	*High*	0.324165436
Parthenium hysterophorus	*Medium*	0.316936045
Lantana camara	*High*	0.313184682
Tithonia diversiflora	*Medium*	0.283112453
Myroxylon balsamum	*Low*	0.278853243
Alstonia macrophylla	*Low*	0.273461163
Imperata cylindrical	*High*	0.265542444
Ziziphus mauritiana	*Medium*	0.262612882
Clidemia hirta	*Medium*	0.255185359
Eichhornia crassipes	*Medium*	0.240426202
Pistia stratiotes	*Medium*	0.240426202
Alternanthera philoxeroides	*Medium*	0.23174
Ulex europaeus	*Medium*	0.216863667
Annona glabra	*Medium*	0.196959
Dillenia suffructicosa	*Low*	0.173985

Table 10.23. Here we present the final aggregated value $DOAGG_i$ as a linguistic label which has been matched with the linguistic scale as in Figure 4.4. This procedure has been carried out using the numerical-linguistic transformation function as defined in Definition 4.3.

After investigating the performance of the model output, the IAPS in the data set has been classified into three categories which pose with different quantifiers. These categories are as follows:

Category I: If a species' $VRS \leq Low$ and $PDS = Unlikely$ and $FDS = Unlikely$ or
$DIS \leq Low$ and $VRS \leq Low$ and $PDS \leq Unlikely$ and $AP = Unlikely$ and $IR =$

Table 10.22

Validation results of Model I – Approximation Approach. (Adapted from H.O.W. Peiris, S.S.N. Perera, S. Chakraverty and S.M.W. Ranwala, "Development of a mathematical model to evaluate the rate of aggregate risk of invasive alien species: Fuzzy risk assessment approach," *International Journal of Bio Mathematics,* 11, no(4) (2018): 1-17.)

Category of Species	Species	NRA	Model I
Invasive	*Austroeupatorium inulifolium*	*High*	0.489959213
	Panicum maximum	*High*	0.470406151
	Cuscuta campestris	*High*	0.351996169
	Pueraria montana	*Medium*	0.23217039
	Acacia mearnsii	*High*	0.549982137
	Myrica faya	*Low*	0.229461789
Non-Invasive	*Cassia fistula*	*Low*	0.191124
	Cissus rotundifolia	*Low*	0.195174
	Hedychium gardnerianum	*Low*	0.180057937
	Magnefera indica	*Low*	0.172286499

Table 10.23

Weights (W) for main and sub parameters. (Adapted from H.O.W. Peiris, S.S.N. Perera, S. Chakraverty and S.M.W. Ranwala, "Development of a mathematical model to evaluate the rate of aggregate risk of invasive alien species: Fuzzy risk assessment approach," *International Journal of Bio Mathematics,* 11, no(4) (2018): 1-17.)

Main parameter	W_{Main}	Sub parameter	W_{Sub}
DIS	*Very High*		
G	-	*VRS*	*Very High*
		FDS	*High*
		PDS	*Very High*
SGL	No weight assigned		
AP	*Very High*		
IR	*Very High*		
MIS	-	*HA*	*High*
		NMD	*High*

$Unlikely$,
then use Model I. Here weights (W) do not assign for performance values of main and sub parameters.

Category II: If a species' $VRS \geqslant Medium$ and $PDS = Unlikely$ and $FDS = High$ and $AP = High$ and $IR = Unlikely$,

then performance values for *VRS, FDS, NMD, DIS, IR* aggregate with W as in

Table 10.24

Test results of Model II – Approximation Approach. (Adapted from H.O.W. Peiris, S.S.N. Perera, S. Chakraverty and S.M.W. Ranwala, "Development of a mathematical model to evaluate the rate of aggregate risk of invasive alien species: Fuzzy risk assessment approach," *International Journal of Bio Mathematics*, 11, no(4) (2018): 1-17.)

Invasive Species	Risk Level (NRA)	Aggregated Risk (numerical value)	Linguistic Labels of Aggregated Risk
Alternanthera philoxeroides	*Medium*	0.525779	*Medium*
Clidemia hirta	*Medium*	0.541375	*Medium*
Miconia calvescens	*High*	0.596138	*High*
Alstonia macrophylla	*Low*	0.273461	*Low*
Annona glabra	*Medium*	0.477566	*Medium*
Clusia rosea	*Low*	0.326705	*Low*
Dillenia suffructicosa	*Low*	0.445421	*Medium*
Ageratina riparia	*Medium*	0.610195	*High*
Mimosa invisa	*High*	0.604604	*High*
Myroxylon balsamum	*Low*	0.278853	*Low*
Tithonia diversiflora	*Medium*	0.588769	*High*
Mikania micrantha	*High*	0.700198	*High*
Prosopis juliflora	*High*	0.62734	*High*
Ulex europaeus	*Medium*	0.496612	*Medium*
Mimosa pigra	*Very High*	0.680319	*High*
Chromolaena odorata	*High*	0.700198	*High*
Parthenium hysterophorus	*Medium*	0.578769	*Medium*
Lantana camara	*High*	0.577937	*Medium*
Imperata cylindrical	*High*	0.587697	*High*
Opuntia stricta	*High*	0.415049	*High*
Colubrina asiatica	*Medium*	0.595311	*High*
Pennisetum polystachion	*High*	0.608643	*High*
Sphagneticola trilobata	*Medium*	0.663411	*High*
Ziziphus marutiana	*Medium*	0.592565	*Medium*
Eichhornia crassipes	*Medium*	0.543293	*Medium*
Pistia stratiotes	*Medium*	0.534714	*Medium*
Leucaena leucocephala	*High*	0.534714	*High*

Table 10.23 with the mean quantifier guide.

Category III: If a species does not belong to either above category, then performance values for *VRS, FDS, NMD, DIS, IR, AP* aggregate with *W* as in Table 10.23 with the mean quantifier guide.

This model has been validated using the same set of species as used in Model I. The test and validation results are tabulated in Tables 10.24 and 10.25.

Table 10.25

Validation results of Model II – Approximation Approach. (Adapted from H.O.W. Peiris, S.S.N. Perera, S. Chakraverty and S.M.W. Ranwala, "Development of a mathematical model to evaluate the rate of aggregate risk of invasive alien species: Fuzzy risk assessment approach," *International Journal of Bio Mathematics,* 11, no(4) (2018): 1-17.)

Category of Species	Species	Risk Level (NRA)	Aggregated Risk	Linguistic Labels of Aggregated Risk
Invasive	*Austroeupatorium inulifolium*	*High*	0.627487	*High*
	Panicum maximum	*High*	0.636772	*High*
	Cuscuta campestris	*High*	0.63324	*High*
	Pueraria montana	*Medium*	0.570898	*Medium*
	Acacia mearnsii	*High*	0.702725	*High*
	Myrica faya	*Low*	0.191124	*Medium*
Non-Invasive	*Cassia fistula*	*Low*	0.195174	*Very Low*
	Cissus rotundifolia	*Low*	0.180058	*Very Low*
	Hedychium gardnerianum	*Low*	0.542407	*Very Low*
	Magnefera indica	*Low*	0.172286	*Very Low*

10.5.1 DISCUSSION – MODEL II – APPROXIMATION APPROACH

According to Table 10.24, it can clearly be seen that risk levels of most species are compatible with NRA risk levels. Also it can be observed that the discrimination occurred in the NRA risk level and the ranges of output of Model I have been clearly solved in Model II.

This observation can be further proven by investigating the validation results tabulated in Table 10.25. It shows that the species in the non-invasive category belong to the *Very Low* risk level as NRA gives the "*Low*" linguistic label.

11 Evaluation of Invasion Risk Levels with Asymmetrically Distributed Linguistic Terms Sets

11.1 AN OVERVIEW

In Chapter 7, we introduced to factor-based models based on the fuzzy 2-tuple approach with an unbalanced linguistic scale. In this chapter, we basically focus on constructing models to evaluate the risk of IAPS by considering an asymmetrically (unbalanced) distributed linguistic terms set for obtaining the performance values and weights of the parameters.

In the data set, performance values and the weights of the parameter for each species have been given from a symmetrically distributed linguistic scale. Therefore, we assume Figure 11.1 as the unbalanced scale for evaluating the performance and weights of the parameters. Figure. 11.1 depicts seven labels with different semantics compared to Figure 4.4.

The first task is to assign the semantics of labels in the unbalanced scale. For that, we have followed the procedure given in Section 7.2. First we start with inputs as described in Figure 7.2.

Inputs

- Unbalanced linguistic terms set S
  The unbalanced scale S is depicted in Fig 11.1. The labels *Unlikely*, *Very Low*, *Low*, *Medium*, *High*, *Very High*, and *Extremely High* are denoted by U, VL, L,M, H, VH, EH respectively.

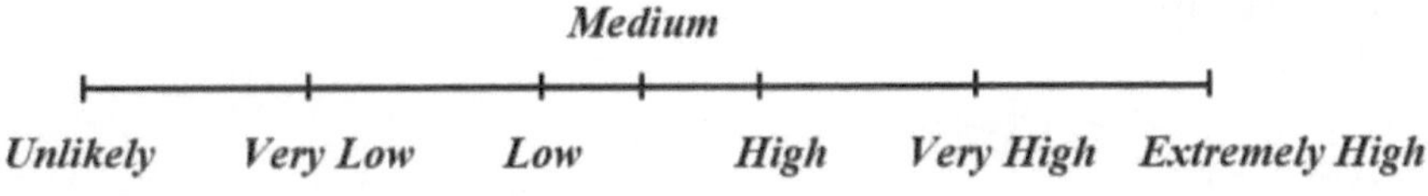

Figure 11.1 Unbalanced linguistic terms set S.

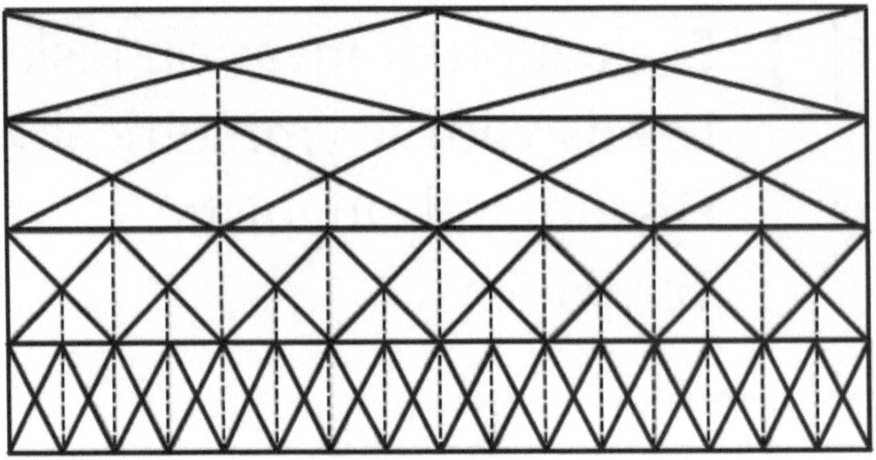

Figure 11.2 Four-level linguistic hierarchy. Reprinted from A. Cid-López, M.J. Hornos, R.A. Carrasco, E. Herrera-Viedma, and F. Chiclana, "Linguistic multicriteria decision-making model with output variable expressive richness," *Expert Systems with Applications* 83 (2017): 350-362, Copyright (2017), with permission from Elsevier.

- Density distribution of S
  According to the scale given in Fig.11.1.
  $S_L = \{U, VL, L\}$
  $S_C = \{M\}$
  $S_R = \{H, VH, EH\}$
  where S_L, S_C, and S_R denote the sets left lateral, central, right lateral respectively.
  Using Eq. (7.1) density distribution may be obtained as
  $\{3, middle\}, 1, \{3, middle\}$
- Linguistic Hierarchy (LH)
  Here we have chosen four levels of hierarchy, which are depicted in Figure 11.2. Here $t = 1, 2, 3, 4$ with cardinality $n(t)$; 2, 5, 9, 17 respectively.

Representation of unbalanced terms of one level of LH Let us take the set S_R. Now check the condition given in Eq. (7.4).

It is clear that

$$\frac{3-1}{2} = 1 \neq \#(S_R)$$
$$\frac{5-1}{2} = 2 \neq \#(S_R)$$
$$\frac{9-1}{2} = 4 \neq \#(S_R)$$
$$\frac{17-1}{2} = 8 \neq \#(S_R)$$

Thus the condition is not satisfied for the set S_R. Similarly we can show that the condition is not satisfied for the set S_L.

Since all the conditions are not satisfied, we need to move to the next step.

Representation of unbalanced terms of two levels of ***LH*** Here two levels t and $t+1$ have to be found such that
$((n(t)-1)/2) < \#(S_R) < ((n(t+1)-1)/2)$.

This shows that $2 < \#(S_R) < 4$. Hence the condition is satisfied when $t=2$ and $t+1=3$.

For the set S_L, the levels t and $t+1$ are similar as in S_R i.e. $t=2$ and $t+1=3$. Let us now move to the next step.

Representation of labels S_R/S_L **from assignable sets and other steps.** First let us start with the set S_R.

- Representation of S_R.

  First divide the set into two subsets such as S_{RE} and S_{RC} where $S_R = S_{RE} \bigcup S_{RC}$. Here the subset S_{RC} consists of the labels close to S_C, and S_{RE} consists of labels close to the Maximum label in S, i.e. label EH.
  Since $density_R$ is middle, the subset S_{RE} is represented by $AS_R^n(t)$, i.e. AS_R^5. Here $Nlab_t = 4-3 = 1$ which indicates the number of highest labels in the subset S_{RE}. Since $Nlab_t + Nlab_{t+1} = 3$ i.e. $t+1, Nlab_{t+1} = 2$.
  The set $AS_R^{n(t)}$ begins with $s_{C+1+\delta}^{n(t+1)}$ where $\delta = round((lab_{t+1}/2))$. Here $\delta = 1$.
  Then $AS_R^5 = \{s_4^5\}$.

  Since $density_R$ is middle, the subset S_{RC} is represented by $AS_R^{n(t+1)}$, i.e. AS_R^9. Here $Nlab_{t+1} = 2$ which indicates the number of smallest labels in the subset S_{RC}.

  The set $AS_R^{n(t+1)}$ begins with the label following the label $S_{C+1}^{n(t+1)}$, i.e. S_4^9. Thus $AS_R^9 = \{s_5^9, s_6^9, s_7^9, s_8^9\}$.

  Now apply the R_{Rep} rule. First take the highest label in the set AS_R^5 i.e. $s_4^5 \leftarrow EH$. One may observe that s_4^5 is in the form of $s_{2\cdot 2}^5$ but there is no associated label in set AS_R^9. Therefore, none of the labels in AS_R^9 needs to be eliminated. Now we may move to the subset S_{RC}.

  Here start with the smallest label, i.e. $H(High)$. Thus $s_5^9 \leftarrow H$. Now take the remaining label VH. Then $s_6^9 \leftarrow H$. According to Figure 11.3, it can be seen that VH is a bridge label since Brid(VH)=True. Therefore the

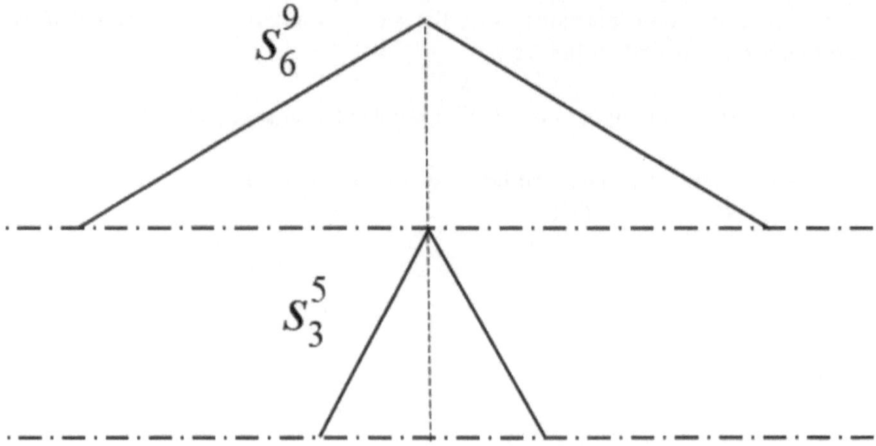

Figure 11.3 Two-level representation of Label *Very High* (VH).

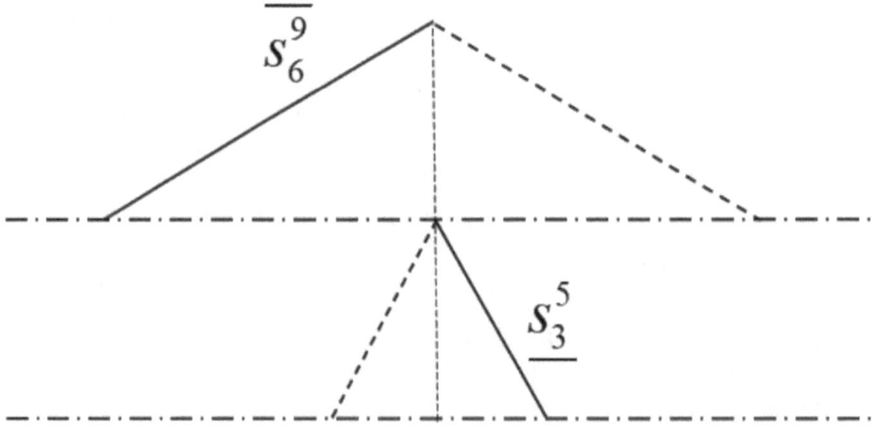

Figure 11.4 Bridge label *Very High* (VH).

representation of VH may be obtained by the labels $\overline{s_6^9}$ and $\underline{s_3^5}$, i.e. $\overline{s_6^9} \cup \underline{s_3^5} \leftarrow VH$ (see Figure 11.4).

Now let us move to the set S_C.

- Representation of S_C

  Here the downside of central label M is $\underline{s_2^5}$. And the upside of central label is $\overline{s_4^9}$. Thus the representation of M is $\overline{s_4^9} \cup \underline{s_2^5}$.

- Representation of S_L

  First divide the set into two subsets such as S_{LE} and S_{LC} where $S_L = S_{LE} \bigcup S_{LC}$. Here the subset S_{LC} consists of the labels close to S_C, i.e. $\{L\}$ and S_{LE} which consists of labels close to the Minimum label in S i.e. $\{U, VL\}$.

  Since $density_L$ is middle, the subset S_{LE} is represented by $AS_R^n(t)$, i.e. AS_R^5. Here $Nlab_{t+1} = 2$ which indicates the number of smallest labels in the subset S_{LE}.
  The set AS_L^5 begins with $s_{C-1-\delta}^5$ where $\delta = round\,((lab_{t+1}/2)) = 1$.
  Then $AS_L^5 = \{s_0^5, s_1^5\}$.

  Since $density_L$ is middle, the subset S_{LC} is represented by $AS_R^{n(t+1)}$, i.e. AS_R^9. Here $Nlab_t = 1$ which indicates the number of largest labels in the subset S_{LC}.

  The set AS_L^9 begins with the label previous to the middle label, i.e. s_3^9.
  Thus $AS_L^9 = \{s_3^9, s_4^9, s_5^9, s_6^9, s_7^9, s_8^9\}$.

  By applying R_{Rep} we get $s_0^5 \leftarrow U$ and there is no associated labels in level 3. Now take the last label VL which is assigned with s_1^5. According to Figure 11.5, it can be seen that VL is a bridge label such that Brid(VL)=True. Therefore the representation of VL has obtained from two levels and as such $\overline{s_1^5} \bigcup \underline{s_2^9}$.

  Now let us move to the label L which is the only element in the set S_{LC}. This label is represented as $s_3^9 \leftarrow L$ which does not have any associated labels.

Let us see the required outputs as follows:
The linguistic terms set S is

$$S = \left\{ s_0^5 \leftarrow U, \overline{s_1^5} \bigcup \underline{s_2^9} \leftarrow VL, s_3^9 \leftarrow L, \overline{s_4^9} \bigcup \underline{s_2^5} \leftarrow M, s_5^9 \leftarrow H, \overline{s_6^9} \bigcup \underline{s_3^5} \leftarrow VH, s_4^5 \leftarrow EH \right\}.$$

The semantic representation of $LH(S)$ is presented in Table 11.1.
And the subsets S_{LE}, S_{LC}, S_C, S_{RC}, S_{RE} are ordered in increasing order.

The set of levels of LH, i.e. $T_{LH} = \{2, 3, 3, 2\}$.
In the next section, we present the models which have been developed based on the unbalanced scale obtained above.

11.2 MODEL I - UNBALANCED

In this section, we produce a model to evaluate the risk of IAPS by following the model structure given in Figure 7.4. We have assumed that the selected parameters are equally important towards invasiveness. Here the parameters as in Table 10.1 have been considered as the model parameters.

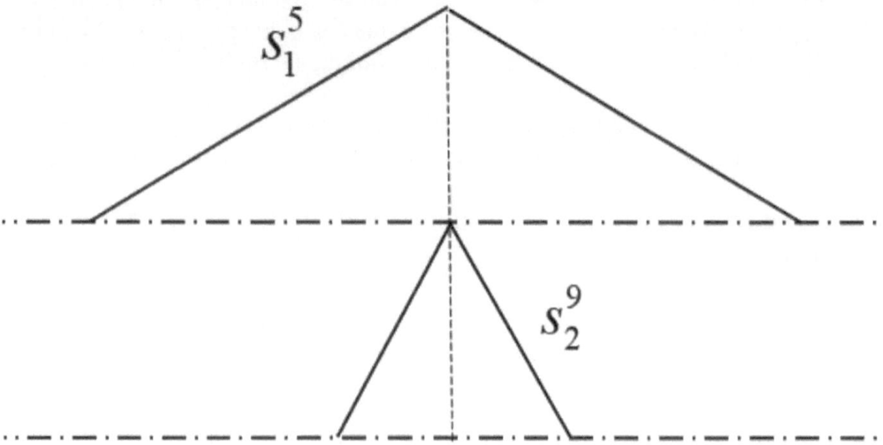

Figure 11.5 Two-level representation of label *Very Low* (*VL*).

As to the model structure, the first three steps have already been accomplished. Therefore, we start with the fourth, i.e. converting the performance values of the parameters into 2-tuples. This has been done by following the procedure given in Section 10.3. The following sections elaborate upon the aggregation of the model parameters.

11.2.1 AGGREGATION OF SUB PARAMETERS

Since the main parameter *MIS* has two sub parameters such as *HA* and *NMD*, these two have to be aggregated first. The aggregation procedure connected with the remaining steps of the model are discussed using an example below:

Consider the species *Prosopis juliflora*. The performance values of the species w.r.t. sub parameters *HA* and *NMD* are $(H,0)$ and $(M,0)$ respectively.

Let us first transform the performance values into *LH* as follows:
$LH(M,0) = s_4^9$
$LH(H,0) = s_5^9$
Here $T_{LH} = 3$

Table 11.1
LH_S and Brid(S)

S	LH_S	Brid(S)
$s_0 = U$	$S_{I(0)}^{G(0)} = s_0^5$	False
$s_1 = VL$	$S_{I(1)}^{G(1)} = s_1^5$	True
$s_2 = L$	$S_{I(2)}^{G(2)} = s_3^9$	False
$s_3 = M$	$S_{I(3)}^{G(3)} = s_4^9$	True
$s_4 = H$	$S_{I(4)}^{G(4)} = s_5^9$	False
$s_5 = VH$	$S_{I(5)}^{G(5)} = s_6^9$	True
$s_6 = EH$	$S_{I(6)}^{G(6)} = s_4^5$	False

Using Definition 7.1 the aggregation is carried out as

$\Lambda^F\left[(M,0),(H,0)\right]$

$=LH^{-1}\left(Agg\left(TF_3^3\left(s_4^9\right),TF_3^3\left(s_5^9\right)\right)\right)$

Since Brid(s_5^9)=False, $TF_3^3\left(s_5^9\right) = TF_3^3\left(s_5^9\right)$.

On the contrary, Brid(s_4^9)=True. Since symbolic translation of M is zero $TF_3^3\left(s_4^9\right) = TF_3^3\left(s_4^9\right)$ (see Case 2.3 in Section 7.3).

Hence
$LH^{-1}\left(Agg\left(TF_3^3\left(s_4^9\right),TF_3^3\left(s_5^9\right)\right)\right)$

$$=LH^{-1}\left(\Delta\left(\frac{1}{2}\left(\Delta^{-1}\left(s_4^9\right)+\Delta^{-1}\left(s_5^9\right)\right)\right)\right)$$

$$=LH^{-1}\left(\Delta\left(\frac{1}{2}(4+5)\right)\right) = LH^{-1}\left(\Delta(4.5)\right)$$

$$=LH^{-1}\left(s_5^9,-0.5\right).$$

Next, we need to find the label in $LH(S)$ which represents the same information as $\left(s_5^9,-0.5\right)$.
For that check the condition given in Eq. (7.5). From Table 11.1, it can be seen that the label "H" is satisfied with the condition given in Eq. (7.5).
Since the label "H" is not a bridge label, this evaluation is related to Case (1.1), step 5 in Section 7.3. Thus $\lambda = \alpha_k = -0.5$.

Then we get $LH^{-1}\left(s_5^9,-0.5\right) = (H,-0.5)$.

In the following, we discuss the aggregation procedure of the main parameters.

11.2.2 AGGREGATION OF MAIN PARAMETERS

In this task, there are only four main parameters such as *DIS*, *VRS*, *SGL*, and *MIS* to be aggregated. This procedure is explained below:

For example, consider the species *Clidemia hirta* with performance values $((H,0),(VL,0),(M,0),(H,-0.5))$ corresponding to parameters *DIS*, *VRS*, *SGL*, *MIS* respectively.

For these values we get:

$$LH(VL,0) = (s_1^5, 0)$$

$$LH(M,0) = (s_4^9, 0)$$

$$LH(H,0) = (s_5^9, 0)$$

$$LH(H,-0.5) = (s_5^9, -0.5)$$

and $t_{Gmax} = 3$.

Observe that Brid(H) is False. Then $TF_3^3(s_5^9,0) = TF_3^3(s_5^9,0)$ from step 6 in Case 1, in Section 7.3.

Further, one may note that Brid(VL)= Brid(M)=True.

Then using Eq. (7.2) we get

$$TF_3^2(s_1^5,0) = \Delta_3\left(\frac{\Delta_2^{-1}(s_1^5)\cdot(9-1)}{(5-1)}\right)$$
$$=\Delta_3 = 2$$
$$=TF_3^3(s_2^9,0).$$

It is clear that $TF_3^3(s_4^9,0)$=$TF_3^3(s_4^9,0)$ since $tLC = t_{RC}$ and $M \in S_C$. Also $TF_3^3(s_5^9,0) = TF_3^3(s_5^9,0)$ and $TF_3^3(s_5^9,-0.5) = TF_3^3(s_5^9,-0.5)$ since Brid(H)=False.

Then we get

$\Lambda^F[(H,0),(VL,0),(M,0),(H,-0.5)]$

$$=LH^{-1}\left(Agg\left(TF_3^3(s_5^9), TF_3^3(s_2^9), TF_3^3(s_4^9), TF_3^3(s_5^9,-0.5)\right)\right)$$
$$=LH^{-1}\left(\Delta\left(\frac{1}{4}(5+2+4+4.5)\right)\right)$$
$$= LH^{-1}(\Delta(3.875))$$
$$=LH^{-1}(s_4^9,-0.125).$$

Table 11.2
Test results of Model I - Unbalanced

Invasive Species	Model	NRA Value
Alternanthera philoxeroides	$(H,-0.375)$	H
Clidemia hirta	$(M,-0.125)$	H
Miconia calvescens	$(H,0)$	H
Alstonia macrophylla	$(M,0.125)$	M
Annona glabra	$(M,-0.5)$	M
Clusia rosea	$(L,0.375)$	M
Dillenia suffructicosa	$(M,-0.375)$	M
Ageratina riparia	$(H,-0.375)$	M
Mimosa invisa	$(M,0)$	H
Myroxylon balsamum	$(M,-0.375)$	M
Tithonia diversiflora	$(M,-0.25)$	M
Mikania micrantha	$(H,-0.25)$	H
Prosopis juliflora	$(M,0.375)$	H
Ulex europaeus	$(L,0.375)$	M
Mimosa pigra	$(M,-0.125)$	H
Chromolaena odorata	$(H,-0.5)$	H
Parthenium hysterophorus	$(M,0.375)$	H
Lantana camara	$(M,0)$	M
Imperata cylindrical	$(H,0.125)$	VH
Opuntia stricta	$(M,0.25)$	H
Colubrina asiatica	$(L,0.375)$	M
Pennisetum polystachion	$(M,-0.375)$	M
Sphagneticola trilobata	$(M,-0.375)$	M
Zizphus mauritiana	$(L,0.125)$	L
Eichhornia crassipes	$(H,-0.5)$	H
Pistia stratiotes	$(H,-0.5)$	H
Leucaena leucocephala	$(M,-0.125)$	M

One may note that the condition given in Eq. (7.5) is satisfied with label "M." Since $t_{LC} = t_{RC}$ and $M \in S_C$, $\lambda = \alpha_k = -0.125$.

Thus $LH^{-1}\left(s_4^9, -0.125\right)$=$(M, -0.125)$.

Following the above mentioned procedure, the risks of 27 IAPS in the data set have been evaluated. Further, a model has been validated using the same set of known invasive and non-invasive species used in the previous models. The test and validation results are presented in Tables 11.2 and 11.3.

11.2.3 DISCUSSION MODEL I - UNBALANCED

According to Table 11.2, except the species *Miconia calvescens* and *Lantana camara*, the risk levels are more or less compared to the corresponding NRA values. Specifically, for species like *Ageratina riparia, Prosopis juliflora, Parthenium hysterophorus, Opuntia stricta*, their risk levels are in between the labels "Medium" and "High", where in most cases the absolute value of α_k is 0.375. Apart from that,

Table 11.3

Validation results of Model I - Unbalanced

Category of Species	Species	NRA	Model
Invasive	*Austroeupatorium inulifolium*	*M*	(*M*,-0.125)
	Panicum maximum	*H*	(*M*,0.125)
	Cuscuta campestris	*H*	(*H*,-0.5)
	Pueraria montana	*H*	(*M*,0)
	Acacia mearnsii	*M*	(*M*,0)
	Myrica faya	*M*	(*M*,-0.375)
	Cassia fistula	*M*	(*M*,-0.375)
Non-Invasive	*Cissus rotundifolia*	*L*	(*L*,0.25)
	Hedychium gardnerianum	*L*	(*L*,-0.25)
	Magnefera indica	*M*	(*L*,0.375)

Clusia rosea, Ulex europaeus, Colubrina asiatica are the species for which those risk levels are in between the labels, "Low" and "Medium" and 0.625 distant from the NRA level.

The validation results in Table 11.3 show that the species in the invasive category show a similar pattern as test results. Moreover, the risk levels of non-invasive species stand behind the label "Medium" as some species take "Medium" as NRA value.

In the next section, we discuss the model which incorporates the important weights of the parameters in a unbalanced linguistic scale.

11.3 MODEL II - UNBALANCED

In this model we incorporate the importance weights of sub and main parameters as presented in Table 10.4.

First the sub parameters of *MIS* have been weights as follows:

Consider the performance, weights, and weighted performance values of the sub parameters *HA* and *NMD* of the species *Miconia calvescens* given in Table 11.4.

Table 11.4

Weighted performance values *HA* and *NMD* of species *Miconia calvescens*

	MIS	
	HA	*NMD*
Importance weight	$(H,0)$	$(M,0)$
Performance value	$(M,0)$	$(M,0)$
Weighted performance value	$(H,-0.5)$	$(M,0)$

Table 11.5
Weighted performance values of species *Mikania micrantha*

	DIS	*VRS*	*SGL*	*MIS*
Importance weight	$(VH,0)$	$(VH,0)$	$(M,0)$	$(H,0)$
Performance value	$(M,0)$	$(VL,0)$	$(M,0)$	$(M,0.25)$
Weighted performance value	$(VH,-0.5)$	$(H,0)$	$(H,-0.5)$	$(H,-0.25)$

Here the importance weight and performance value have to be aggregated using a similar procedure as in the previous section. Follow the first example given in Subsection 11.2.1. Then using the same procedure, the weighted performance values such as $(H,-0.5)$ and $(M,0)$ have to be aggregated as follows:

Here $LH(H,-0.5) = (s_5^9,-0.5)$

$LH(M,0) = (s_4^9,0)$.

One may note that

$TF_3^3(s_5^9,-0.5) = TF_3^3(s_5^9,-0.5)$ since Brid(H)=False.
$TF_3^3(s_4^9,0)$=$TF_3^3(s_4^9,0)$ since $tLC = t_{RC}$ and $M \in S_C$.

Then we get
$\Lambda^F[(H,-0.5),(M,0)]$

$$=LH^{-1}(Agg(TF_3^3(s_5^9,-0.5),TF_3^3(s_4^9,0)))$$
$$=LH^{-1}\left(\Delta\left(\frac{1}{2}(4.5+4)\right)\right)$$
$$=LH^{-1}(\Delta(4.25))$$
$$=LH^{-1}(s_4^9,0.25).$$

Observe that the condition given in Eq. (7.5) is satisfied with the label "M." Since $tLC = t_{RC}$ and $M \in S_C$, $\lambda = \alpha_k = 0.25$.

Thus $LH^{-1}(s_4^9,0.25)$=$(M,0.25)$.

Therefore, the weighted performance value of the main parameters MIS of *Miconia calvescens* is $(M,0.25)$

Likewise, the same procedure has been applied to evaluate the weighted performance values of the main four parameters. Table 11.5 presents the weighted performance values of main parameters of the species *Mikania micrantha*.

The final aggregated value of four weighted main parameters is obtained as follows:

$LH(H,0) = (s_5^9,0)$

Table 11.6

Test results of Model II - Unbalanced

Invasive Species	Model	NRA Value
Alternanthera philoxeroides	(*H*,-0.0625)	*H*
Clidemia hirta	(*H*,-0.4375)	*H*
Miconia calvescens	(*H*,0.1875)	*H*
Alstonia macrophylla	(*H*,-0.25)	*M*
Annona glabra	(*H*,0.4375)	*M*
Clusia rosea	(*H*,0.375)	*M*
Dillenia suffructicosa	(*H*,-0.5)	*M*
Ageratina riparia	(*H*,0)	*M*
Mimosa invisa	(*H*,0)	*H*
Myroxylon balsamum	(*H*,-0.5)	*M*
Tithonia diversiflora	(*H*, -0.4375)	*M*
Mikania micrantha	(*H*, -0.0625)	*H*
Prosopis juliflora	(*H*,-0.1875)	*H*
Ulex europaeus	(*H*,0.4375)	*M*
Mimosa pigra	(*H*,-0.4375)	*H*
Chromolaena odorata	(*H*,-0.0625)	*H*
Parthenium hysterophorus	(*H*,-0.1875)	*H*
Lantana camara	(*H*,-0.3125)	*M*
Imperata cylindrical	(*H*,0.1875)	*VH*
Opuntia stricta	(*H*,-0.1875)	*H*
Colubrina asiatica	(*M*,0.375)	*M*
Pennisetum polystachion	(*M*,0.4375)	*M*
Sphagneticola trilobata	(*M*,0.4375)	*M*
Zizphus mauritiana	(*M*,0.25)	*L*
Eichhornia crassipes	(*H*,-0.0625)	*H*
Pistia stratiotes	(*H*,-0.0625)	*H*
Leucaena leucocephala	(*H*,-0.375)	*M*

$$LH(M,0) = (s_4^9, 0)$$

$$LH(H,-0.25) = (s_5^9, -0.25).$$

Note that

$TF_3^3(s_6^9, -0.5) = TF_3^3(s_6^9, -0.5)$ since Brid(VH)=True and $\alpha_k \leq 0$.
$TF_3^3(s_5^9, 0) = TF_3^3(s_5^9, 0)$,
$TF_3^3(s_5^9, -0.5) = TF_3^3(s_5^9, -0.5)$,
$TF_3^3(s_5^9, -0.25) = TF_3^3(s_5^9, -0.25)$
since Brid(H)=False.

Then we get
$\Lambda^F[(VH,-0.5),(H,0),(H,-0.5),(H,-0.25)]$

Table 11.7
Validation results of Model II - Unbalanced

Category of Species	Species	NRA	Model
Invasive	*Austroeupatorium inulifolium*	*M*	(*H*,-0.4375)
	Panicum maximum	*H*	(*H*,-0.3125)
	Cuscuta campestris	*H*	(*H*,-0.0625)
	Pueraria montana	*H*	(*H*,-0.4531)
	Acacia mearnsii	*M*	(*H*,-0.3125)
	Myrica faya	*M*	(*H*,-0.5)
Non-Invasive	*Cassia fistula*	*M*	(*M*,0.4375)
	Cissus rotundifolia	*L*	(*M*,3125)
	Hedychium gardnerianum	*L*	(*M*,0.0625)
	Magnefera indica	*M*	(*M*,0.3125)

$$=LH^{-1}\left(Agg\left(TF_3^3\left(s_6^9,-0.5\right),TF_3^3\left(s_5^9,0\right),TF_3^3\left(s_5^9,-0.5\right),TF_3^3\left(s_5^9,-0.25\right)\right)\right)$$

$$=LH^{-1}\left(\Delta\left(\frac{1}{4}(5.5+5+4.5+4.75)\right)\right)$$

$$=LH^{-1}\left(\Delta(4.9375)\right)$$

$$=LH^{-1}\left(s_5^9,-0.0625\right).$$

It is clear that the condition given in Eq. (7.5) is satisfied with the label "*H*." Since Brid(*H*)=False, $\lambda=\alpha_k=-0.0625$.

Thus $LH^{-1}\left(s_5^9,-0.0625\right)=(H,-0.0625)$.

Therefore, the risk value of the species *Mikania micrantha* is $(H,-0.0625)$.

Following a similar procedure, the risks of IAPS in the data set have been evaluated and are presented in Table 11.6. Further, the model has been validated and the results are presented in Table 11.7.

11.3.1 DISCUSSION - MODEL II - UNBALANCED

According to Table 11.6, it is observed that, species like *Alternanthera philoxeroides, Clidemia hirta, Mimosa invisa, Mikania micrantha, Prosopis juliflora, Chromolaena odorata, Parthenium hysterophorus, Imperata cylindrical, Opuntia stricta, Colubrina asiatica, Eichhornia crassipes, Pistia stratiotes* have been able to reduce the gap between model output and NRA level compared to Model I. As invasive species in the validation results follow the same pattern.

On the contrary, the non-invasive species show slightly higher values compared to Model I.

Bibliography

Abrerizo, F.J., Al-Hmouz, R., Morfeq, A., et al. 2017. Soft consensus measures in group decision making using unbalanced fuzzy linguistic information. *Appl. Soft Comput.* 21: 3037-3050.

Aloo, P., Ojwang, W., Omondi, R., Njiru, J. M. , and Oyugi, D. 2013. A review of the impacts of invasive aquatic weeds on the biodiversity of some tropical water bodies with special reference to Lake Victoria (Kenya). *Biodivers*, 4(4): 471-18.

Alpert, P., Bone, E., and Holzapfel, C. 2000. Invasiveness, invisibility and the role of environmental stress in the spread of non-native plants. *Perspect. Plant Ecol. Evol. Syst.* 3: 52-66.

Arbačauskas, K., Semenchenko, V., Grabowski, M., Leuven R., Paunović, M., Son, O. et al. 2008. Assessment of biocontamination of benthic macroinvertebrate communities in European inland waterways. *Aquat. Invasions.*, 3(2): 211-230.

Alefeld, G., and Mayer, G. 2000. Interval analysis: theory and pplications. *J. Comput. Appl. Math.* 121: 421-464.

Bai, Y. and Wang, D. 2006. Fundamentals of Fuzzy Logic Control, Fuzzy Sets, Fuzzy Rules and Defuzzifications, In *Advanced Fuzzy Logic Technologies in Industrial Applications*, 17-36. Springer.

Baker, H.G. 1965. Characteristics and modes of origin of weeds. In *The Genetics of Colonizing Species*, eds. H.G. Baker and G.L. Stebbins, 147-172. Academic Press, New York.

Bartomeus, I., Fründ, J., and Williams, N. M. 2015. Invasive plants as novel food resources, the pollinators' perspective, In *Biological Invasions and Behaviour*, eds. D. Sol and J. Weiss, 1-21. Cambridge University Press. (Bartomeus, Fründ, Williams 2015)

Bentbib, A.H. 2002. Solving the full rank interval least squares problem. *Appl. Numer. Math.* 41(2): 283-294.

Berkan, R.C. and Trubatch, S.L. 2000. *Fuzzy System Design Principles*. IEEE press, Delhi.

Bjerknes, A., Totland, Ø., Hegland, S. J., Nielsen, A. 2007. Do alien plant invasions really affect pollination success in native plant species? *Biol. Conserv.* 40: 301-314.

Bojadziev, G. and Bojadziev, M., 2007. *Fuzzy Logic for Business, Finance, and Management*. World Scientific.

Bradley, B.A., Blumenthal, D.M., Wilcove, D.S., Ziska, L.H. 2010. Predicting plant invasions in an era of global change. *Trends Ecol. Evol.* 25: 310-318.

Brooks, M.L. 2008. Plant invasions and fire regimes. In *General Technical Report of USDA Forest Service*, 42(6):33-45.

Cai, M., Gong, Z., and Yu, X. 2017. A method for unbalanced linguistic term sets and its application in group decision making. *Int. J. Fuzzy Syst.* 19(3): 671-682.

CBD.1994. Convention on Biological Diversity; Text and Annexures, Secretariat for Convention on Biological Diversity.

Center, T.D., Frank J. H., et al. 1997. Biological control. In *Strangers in Paradise*, eds. D. Simberloff, D. C. Schmitz, and T. C. Brown, 245-266. Island Press, Washington DC.

Chang, P. and Chen, Y. 1994. A fuzzy multicriteria decision making method for technology transfer strategy selection in biotechnology. *Fuzzy Sets Syst*, 63: 131-139.

Chen, S.J. and Hwang, C.L. 1992. *Fuzzy Multiple Attribute Decision Making–Methods and Applications*. Springer, Berlin.

Chenje, M. and Mohamed-Katerere, J. 2006. Invasive alien species, In *United Nations Environmental Programme (UNEP) African Environment Outlook* 2, 331-349. UNEP, Nairobi, Kenya.

Cid-Lpez, A., Hornos, M.J., Carrasco, R.A. et al. 2017. Linguistic multi-criteria decision-making model with output variable expressive richness. *Expert Syst. Appl.*, 83: 350-362.

Cinar, N. 2009. A decision support model for bank branch location selection. *International Journal of Mechanical, Industrial Science and Engineering* 3: 1-6.

Cronk, Q.U.B. and Fuller, J.L.1995. Plant invaders: The threat to natural ecosystems. In *'People and Plant' Conservation Manual*, London. Chapman and Hall, 1-42

Davis, M.A., Grime, J.P., Thompson, K. 2000. Fluctuating resources in plant communities: a general theory of invasibility. *J Ecol* 88: 528-534.

Delgado, M., Herrera, F., Herrera-Viedma, E., et al. 1998. Combining numerical and linguistic information in group decision making. *J. Inf. Sci.* 107: 177-194.

Delgado, M., Vila, A.M., and Voxman, W. 1998. On a canonical representation of fuzzy numbers. *Fuzzy Set Syst.* 93: 125-135.

Dong, Y. and Herrera-Viedma, E. 2015. Consistency-driven automatic methodology to set interval numerical scales of 2-tuple linguistic term sets and its use in the linguistic GDM with preference relation. *IEEE T Cybernetics* 45(4): 780-792.

Dong, Y., Li, Cong-Cong., and Herrera, F. 2016. Connecting the linguistic hierarchy and the numerical scale for the 2-tuple linguistic model and its use to deal with hesitant unbalanced linguistic information. *Inf. Sci.* 367-368: 259-278.

Dong, Y., Li, CC., Xu, Y., et al. 2015. Consensus-based group decision making under multi-granular unbalanced 2-tuple linguistic preference relations. *Group Decis. Negot* 24: 217-242.

Dubois, D. and Prade, H. 1980a. *Fuzzy and Systems Theory and Applications*. Academic Press, New York.

Dubois, D. and Prade, H. 1980b. New Results about properties and semantics of fuzzy set-Theoretic operators. In *Fuzzy Sets*, eds. P.P. Wang and S.K. Chang, 59-75. Springer, Boston, MA.

El-Keblawy, A. and Al-Rawai, A. 2007. Impacts of the invasive exotic Prosopis juliflora (Sw.) D.C. on the native flora and soils of the UAE. *Plant Ecol*, 190: 23-35.

Ellstrand, N.C and Kristin, A. 2000. Hybridization as a stimulus for the evolution of invasiveness in plants. *Proceedings of National Academy of Sciences USA*, 97:7043-7050.

Fodor, J. and Roubens, M. 1994. *Fuzzy Preference Modelling and Multicriteria Decision Support*, Kluwer Academic Publishers, Dordrecht.

Gabriella, P. and Yager, R.R. 2006. Modeling the concept of majority opinion in group decision making. *Inf. Sci.* 176: 390-414.

Gallardo, B. and Aldridge, D.C. 2013. The dirty dozen: Socio economic factors amplify the invasion potential of 12 high risk aquatic invasive species in Great Britain and Ireland. *J Appl Ecol*, 50: 757-766.

Gooden, B., French, K., Turner, P. J., and Downey, P.O. 2009. Impact threshold for an alien plant invader, Lantana camaraL. on native plant communities. *Biol. Conserv.*, 142: 2631-2641.

Gordon, D.R., Mitterdorfer, B., Pheloung, P.C., Ansari, S., Buddenhagen, C., et al. 2010. Guidance for addressing the Australian Weed Risk Assessment questions. *Plant Prot.Q.*, 25(2): 56-74.

Harris, J. 2006. *Fuzzy Logic Applications in Engineering Science*. Springer.

Herrera, F. and Herrera-Viedma, E. 1997. Aggregation operators for linguistic weighted information. *IEEE Trans. Syst. Man Cybern. Syst.* 27: 646 -656.

Herrera, F. and Herrera-Viedma, E. 2000. Linguistic decision analysis: steps for solving decision problems under linguistic information. *Fuzzy Set Syst.* 115: 67-82.

Herrera, F. and Martinez, L. 2000. A 2-tuple fuzzy linguistic representation model for computing with words. *IEEE T Fuzzy Syst.* 8: 746-752.

Herrera, F., Herrera-Viedma, E., and Martinez, L. 2008. A fuzzy linguistic methodology to deal with unbalanced linguistic term sets. *IEEE T Fuzzy Syst.* 16(2): 354-370.

Herrera, F., Herrera-Viedma, E., and Verdegay, J.L. 1995. A sequential selection process in group decision making with linguistic assessment. *Inf. Sci.* 85: 223-239.

Herrera, F., Herrera-Viedma, E., and Verdegay, J.L. 1996. A model of consensus in group decision making under linguistic assessment. *Fuzzy Set Syst.* 79: 73-87.

Herrera-Viedma, E., Pasi, G., Lopez-Herrera, A.G., and Porcel, C. 2006. Evaluating the information quality of web sites: A methodology based on fuzzy computing with words. JASIST 57: 538-549.

Hiebert, R.D. and Stubbendieck, J., 1993. *Handbook for Ranking Exotic Plants for Management and Control.* US Department of Interior, National Park Service, Denver, Colorado.

Higgins, S.I, Richardson, D.M and Cowling, R.M. 1999. Predicting the landscape-scale distribution of alien plants and their threat to plant diversity. *Conserv. Biol.*, 13(2): 303-313.

Hu, C. 2011. Interval Function and its Linear Least-squares Approximation, Computer Science Department, University of Central Arkansas. http://www.csd.uwo.ca/moreno/SNC-11-filefor-ACM/p16-Hu.pdf. (Accessed 12 October 2015).

Hu, C., De Korvin, R.B.K.A., and Kreinovich, V. 2008. Knowledge Processing with Interval and Soft Computing, In *Advanced Information and Knowledge Processing*, eds. Jain, L., et al. Springer.

Huang H. and Ho, C. (2013) Applying the fuzzy analytic hierarchy process to consumer decision making regarding home stays. *Int J Adv Comput Sci Appl* 5(4): 981-989.

Huynh, V.N., Ho, T.B., and Nakamori, Y. 2002. A parametric representation of linguistic hedges in Zadeh's fuzzy logic. *Int. J Approx. Reason.*, 30: 203-223.

Iqbal, M., Wijesundera, D., and Ranwala, S. 2014. Climate change, invasive alien flora and concerns for their management in Sri Lanka. *Ceylon J. Sci* (Bio. Sci.), 43(2): 1-15.

Jin, Wu, P., Merigo, J.M., and Peng, Z. 2019. Generalized Hamacher aggregation operators for intuitionistic uncertain linguistic sets: Multiple attribute group decision making methods. *Information*, 10(6): 206.

Johns. C.G., Lawton, G.H., Shachak, M. 1994. Organisms as ecosystem engineers. *Oikos* 69: 373-386.

Kacprzyk, J.J. and Fedrizzi, M. 1990. *Multi-person Decision Making Models Using Fuzzy Sets and Possibility Theory*. Kluwer Academic Publishers, Dordrecht.

Kandel, A. 1992. *Fuzzy Expert Systems*. Boca Raton: CRC Press.

Kannan, R., Shackleton, C.M., Krishnan, S. and Shaanker, R.U. 2016. Can local use assist in controlling invasive alien species in tropical forests? The case of Lantana camara in southern India. *For. Ecol. Manag.*, 376:166-173.

Kickert, W.J.M. 1979. *Fuzzy Theories on Decision Making.* Springer.

Kolar, C.S. and Lodge, D.M. 2001. Progress in invasion biology predicting invaders. *Trends Ecol. Evol.*, 16: 199-204.

Kózy L.T. 2014. A Note on Hamacher-Operators. In *Advances in Soft Computing, Intelligent Robotics and Control, Topics in Intelligent Engineering and Informatics*, eds. Fodor, J.,

Fullér, R., 8: 159-163. Springer, Cham.

Koop, A. 2012. Special Applications of Pest Risk Analysis; Weed Risk Assessment. In *Plant Pest Risk Analysis: Concepts and Application*, eds. C. Devorshak. CAB International.

Kotagama, S. W. and Bambardeniya, C.N.B. 2006. An overview of the wetlands of Sri Lanka and their conservation significance, In *National Wetland Directory of Sri Lanka*, 7-16.

Langeland, K.A. 1996. Hydrilla verticillata (L.F.) Royle (Hydrocharitaceae), "The perfect aquatic weed". *Castanea*, 61 (3):293-304.

Lee, H. 1996. Group decision making using fuzzy sets theory for evaluating the rate of aggregative risk in software development. *Fuzzy Set Syst.* 80: 261-271.

Lee, K.H. 2005. *First Course on Fuzzy Theory and Applications*. Springer, Heidelberg

Lee, Li-Wei. and Chen, Shyi-Ming. 2015. Fuzzy decision making based on likelihood-based comparison relations of hesitant fuzzy linguistic term sets and hesitant fuzzy linguistic operators. *Inf. Sci.* 294: 513-529.

Lemaire, J. 1990. Fuzzy Insurance. *Astin Bulletin* 20(1): 33-55.

Li, C-C., Dong, Y., Herrera, F., et al., 2017. Personalized individual semantics in computing with words for supporting linguistic group decision making: An application on consensus reaching.*Inform. Fusion.* 33: 29-40.

Lin, H., Cao, B., and Liao, Y. 2018. *Fuzzy Sets Theory Preliminary*. Springer.

Lindsay, E. A. and French, K. 2006. Litterfall and nitrogen cycling following invasion by Chrysanthemoides monilifera ssp. rotundata in coastal Australia. *J. Appl. Ecol.*, 42(3):556-566.

Liang, W., Zhao, G., and Luo, S. 2018. Linguistic neutrosophic Hamacher aggregation operators and the application in evaluating land reclamation schemes for mines. *PLOS One*, 13:1-29.

Lowen, R. 1992. *Fuzzy Logic: State of the Art*. Kluwer Academic Publishers, Dordrecht.

Mack, R. N., Simberloff, D., Lonsdale, W. M., Evans, H., Clout, M., and Bazzaz, F. A. 2000. Biotic invasions: causes, epidemiology, global consequences, and control. *Ecol Appl*, 10: 689-710.

Marambe, B., Bambaradeniya, C., Pushpakumara, D.K.N.G., and Pallewatta, N. 2001. Human dimensions of invasive alien species in Sri Lanka, In *The Great Reshuffling–Human Dimensions of Invasive Alien Species in Sri Lanka*, ed. J.A. McNeely, 135-139. World Conservation Union, Gland, Switzerland and Cambridge, UK.

Martin, A.J., Nejad, H., Colmar, S.H., and Liem, G.A.D. 2013. Adaptability: How students' responses to uncertainty and novelty predict their academic and non-academic outcomes. *J. Educ. Psychol*, 105(3):728.

Matheworks.2020. (Accessed 02 June 2020.) https://www.mathworks.com/

Medasani, S., Kim, J., and Krishnapuram, R. 1998. An overview of membership function generation techniques for pattern recognition. *Int. J Approx. Reason.* 19: 391-417.

Mendel, J.M. 2006. Computing with words and its relationship with fuzzistics. *Inf. Sci.* 177(4): 988-1006.

Meyerson, L. A. and Reaser, J. K. 2002. Biosecurity: moving toward a comprehensive approach. *BioScience*, 52(7): 593-600.

Morente-Molinera, J,A., Mezei, J., Carlsson, C., and Herrera-Viedma, E. 2017. Improving supervised learning classification methods using multi-granular linguistic modeling and fuzzy entropy. *IEEE T Fuzzy Syst.* 25(5): 1078 - 1089.

Natural Heritage, Virginia Department of Conversation & Recreation. http://www.dcr.virginia.gov. (Accessed 12th July 2013.)

Orlovsky, S.A. 1994. *Calculus of Decomposable Properties, Fuzzy Sets and Decisions*. Allerton Press.

Parker, I.M., Simberloff, D., Lonsdale, W.M., Goodell, K., Wonham, M., Kareiva, P.M., et al.1999. Toward a framework for understanding the ecological effects of invaders. *Biol. Invasions*, 1(1):3-19.

Peiris, H.O.W., Chakraverty, S., Perera, S.S.N., and Ranwala, S.M.W. 2016. Development of a risk assessment mathematical model to evaluate invasion risk of invasive alien species using multivariate linear regression. *Br J Appl Sci Techno* 16(1): 1-11.

Peiris, H.O.W., Chakraverty, S., Perera, S.S.N., and Ranwala, S.M.W. 2017. Novel fuzzy based model on analysis of invasiveness due to dispersal related traits of plants. *Ann. Fuzzy Math. Inform.* 13(3): 6-14.

Peiris, H.O.W., Chakraverty, S., Perera, S.S.N., and Ranwala, S.M.W. 2018. Modelling Dispersal Risk of Invasive Alien Plant Species. In *Recent Advances in Applications of Computational and Fuzzy Mathematics*, eds. S. Chakraverty and S.S.S. Perera, 109-145. Springer.

Peiris, H.O.W., Chakraverty, S., Perera, S.S.N., and Ranwala, S.M.W. 2018. Novel interval multiple linear regression model to assess the risk of invasive alien plant species. *JSC* 9 (1): 12-30.

Peiris, H.O.W., Chakraverty, S., Perera, S.S.N., and Ranwala, S.M.W. 2017. Novel fuzzy linguistic based mathematical model to assess risk of invasive alien plant species. *Appl. Soft Comput.* 59: 326-339.

Peiris H.O.W., Perera S.S.N., Chakraverty S., and Ranwala, S.M.W. 2018. Development of a mathematical model to evaluate the rate of aggregate risk of invasive alien species: Fuzzy risk assessment approach. *Int. J. Biomath.* 11 (4): 1-17.

Pheloung, P.C., Williams, P.A. and Halloy, S.R. 1999. A weed risk assessment model for use as a biosecurity tool evaluating plant introductions. *J. Environ. Manage.*, 57: 239-251.

Pimental, D., Mcnair, S., Janecka, J., Wightman, J., Simmonds, C., and O'Connell, C. 2001. Economic and environmental threats of alien plant, animal, and microbe invasions. *Agric Ecosyst Environ*, 84(1):1.

Pimentel, D., Lach, L., Zuniga, R., and Morrison, D. 2000. Environmental and economic costs of nonindigenous species in the United States. *BioScience*, 50(1):53-65.

Pyšek, P. and Richardson, D.M. 2007. Traits associated with invasiveness in alien plants: Where do we stand? In *Biological Invasions, Ecological Studies*, eds. W. Nentwig, 97-126. 193, Springer Verlag, Berlin & Heidelberg.

Pyšek, P., Jarošík, V., Pergl, J., Randall, R., Chytrý, M., Kühn, I., et al. 2009. The global invasion success of Central European plants is related to distribution characteristics in their native range and species traits. *Divers. Distrib.* 15:891- 903.

Pyšek, P., Jarošík, V. et al. 2012. A global assessment of invasive plant impacts on resident species, communities and ecosystems: the interaction of impact measures, invading species' traits and environment. *Glob Chang Biol*, 18(5): 1725-1737.

Ranwala S.M.W. (2010) Risk Assessment for Invasive Alien Species, In *Invasive Alien Species Strengthening Capacity to Control Introduction and Spread in Sri Lanka*, eds. B. Marambe, P. Silva, S. Wijesundera, N. Attapattu. Biodiversity Secretariat, Ministry of Environment and Natural Resources, Sri Lanka.

Rejmanek, M. and Richardson, M.D. 1996. What attributes make some plants species more invasive. *J. Ecol* 77: 1655-1661.

Richardson, D. M., Macdonald, I. A. W., and Forsyth, G.G. 1989. Reductions in plant species richness under stands of alien trees and shrubs in the Fynbos Biome. *South. Afr. For. J.*,

149:1-8.

Richardson, D.M., Pysek, P., Rejamanek, M., Barbour, M.G., Panetta, F.D., and West, C.J. 2000. Naturalization and invasion of alien plants: Concepts and definitions. *Divers. Distrib*, 6:93-107.

Saaty, T.L. 2008. Decision making with the analytical hierarchy process. *J Serv Sci* 1(1): 83-98.

Seikh, M.R. and Mandal, U. 2019. Intuitionistic fuzzy Dombi aggregation operators and their application to multiple attribute decision-making. *Granul. Comput.*

Shi, L. and Ye, J. 2018. Dombi aggregation operators of neutrosophic cubic sets for multiple attribute decision-making. *Algorithms*, 11(3):29.

Souza-Alonso, P., Novoa, A., and González, L. 2014. Soil biochemical alterations and microbial community responses under Acacia dealbata Link invasion. *Soil Biol. Biochem.*, 79:100-108.

Stockburger, D.W. 1998. Multivariate Statistics: Concepts, Models, and Applications, Southwest Missouri State University.
http://www.psychstat.missouristate.edu/multibook/mlt08m.html (Accessed 10 September 2015).

Stromberg, J.C., Lite, S.J., Marler, R., Paradzick, C., Shafroth, P.B., Shorrock, D., et al. 2007. Altered stream-flow regimes and invasive plant species: the Tamarix case. *Glob. Ecol. Biogeogr.*, 16(3): 381-393.

Tao, Z., Chen, H., Song, X., Zhou, L., and Liu, Jinpei. 2015. Uncertain linguistic fuzzy soft sets and their applications in group decision making.*Appl. Soft Comput.* 34:587-605.

Thomsen, M.S. 2010. Experimental evidence for positive effects of invasive seaweed on native invertebrates via habitat-formation in a seagrass bed. *Aquat. Invasions*, 5(4): 341-346.

Vaidya, O.S. and Kumar, S. 2006. Analytic hierarchy process: An overview of applications. *Eur. J. Oper. Res* 169: 1-29.

Van De Geer, S.A. 2005. Least Squares Estimation, In *Encyclopedia of Statistics in Behavioral Science*, eds. Everitt, B.S. et al., 1041-1045. John Wiley & Sons, Ltd, Chichester.

Veich, C.R. and Clout, M.N. 2002. Turning the tide of biological invasion: the eradication of invasive species. IUCN SSG Invasive Species Specialist Group, IUCN Gland Switzerland and Cambridge, 414.

Vilá, M., Espinar, J.L., et al. 2011. Ecological impacts of invasive alien plants: a meta-analysis of their effects on species, communities and ecosystems. *Ecol. Lett*,14(7): 702-708.

Vitousek, P.M., D'Antonio, C. M., Loope, L.L., Rejmanek, M., and Westbrooks, R. 1997. Introduced species: a significant component of human-caused global change. *N. Z. J. Ecol.*, 21:1-16.

Wang, Y., Muller-Scharer, H., Van Kleunen, M, Cai, A., Zhang, P., Yan, R., et al. 2017. Invasive alien plants benefit more from clonal integration in heterogeneous environments than natives. *New Phytol.*, 216:1072-1078.

Wei, G., Alsaadi, F.E., Hayat, T., et al. 2018. Picture 2-tuple linguistic aggregation operators in multiple attribute decision making. *Appl. Soft Comput.* 22: 989-1002.

Wickramathilake, B. A. K., Weerasinghe, T. K., Ranwala, S.M.W. 2013. Impacts of woody invader Dillenia suffruticosa (Griff.) Martelli. on physico-chemical properties of soil and below and above ground flora. *JTFE* 3(2): 66-75.

Yager, R.R. 1980a. On a general class of fuzzy connectives. *Fuzzy Set Syst.*, 4: 235-242.

Yager, R.R. 1980b. On ordered weighted averaging aggregation operators in multicriteria decision making. *IEEE Trans. Syst. Man Cybern. Syst.* 18: 183-190.

Yager, R.R. 2001. The power average operator. *IEEE Trans. Syst. Man Cybern. Syst.* 31: 724 - 731.

Yager, R.R., Goldstein, L.S., and Mendels, E. 1994. FUZMAR: an approach to aggregating market research data based on fuzzy reasoning. *Fuzzy Sets Syst*, 68: 1-11.

Yager, R.R. and Zadeh, L.A. 1992. *An Introduction to Fuzzy Logic Applications in Intelligent Systems*. Kluwer Academic Publishers, Dordrecht.

Yu, W., Zhang, Z., Zhong, Q., et al. 2017 Extended TODIM for multi-criteria group decision making based on unbalanced hesitant fuzzy linguistic term sets. *Comput. Ind. Eng.* 114: 316-328.

Zadeh, L.A. 1972. A fuzzy-set-theoretic interpretation of linguistic hedges. *Cybern. Syst* 2: 4-34.

Zadeh, L.A. 1975. The concept of a linguistic variable and its applications to approximate reasoning. *Inf. Sci.* 8: 199-249.

Zaghloul, A., Saber, M., Gadow, S., and Awad, F.2020. Biological indicators for pollution detection in terrestrial and aquatic ecosystems. *Bulletin of National Research Center*, 44: 127.

Zhang. L. 2018. The Hamacher aggregation operators and their application to decision making with hesitant pythagorean fuzzy sets. *Advances in Intelligent Systems Research*, 163: 597-601.

Zhang, Y., Xu, Z., Wang, H., and Liao, H. 2016. Consistency-based risk assessment with probabilistic linguistic preference relation. *Appl. Soft Comput.* 49: 817-833.

Zhao J. and Bose, K.B. 2002. Evaluation of Membership Functions for Fuzzy Logic controlled induction motor drive. IEEE 2002 28th Annual Conference of the Industrial Electronics Society, 1: 229-234.

Zimmermann, H.J. 2001. *Fuzzy Set Theory and Its Applications*. Springer (India) Private Limited,New Delhi.

Index